AF410707

Novel Processing Technology of Dairy Products

Novel Processing Technology
of Dairy Products

Editor

Ekaterini Moschopoulou

MDPI • Basel • Beijing • Wuhan • Barcelona • Belgrade • Manchester • Tokyo • Cluj • Tianjin

Editor
Ekaterini Moschopoulou
Agricultural University of
Athens
Greece

Editorial Office
MDPI
St. Alban-Anlage 66
4052 Basel, Switzerland

This is a reprint of articles from the Special Issue published online in the open access journal *Foods* (ISSN 2304-8158) (available at: https://www.mdpi.com/journal/foods/special_issues/Nove-Processing_Technology_Dairy_Products).

For citation purposes, cite each article independently as indicated on the article page online and as indicated below:

LastName, A.A.; LastName, B.B.; LastName, C.C. Article Title. *Journal Name* **Year**, *Volume Number*, Page Range.

ISBN 978-3-0365-4583-7 (Hbk)
ISBN 978-3-0365-4584-4 (PDF)

Contents

About the Editor

Ekaterini Moschopoulou

Ekaterini Moschopoulou (Assistant Professor) has received a B.Sc. degree from the Agricultural University of Athens (AUA), M.Sc. degree in Dairy Science from Reading University of U.K., and Ph.D. degree from AUA. In 2009, she was promoted to a Lecturer position in the Laboratory of Dairy Research of the Department of Food Science & Human Nutrition of AUA and in 2018 to the Assistant Professor position. She teaches theoretical and lab courses of Dairy Technology to graduate and postgraduate students. Her research interests include milk clotting enzymes; cheese, yoghurt and ice cream technology; membrane technology; the implementation of modern processing methods in milk and dairy products; the development of new dairy products; milk heat treatment; and the valorization of whey/acid whey. She is a peer-reviewer and guest editor in scientific journals in the area of Dairy Science and Technology.

 foods

MDPI

Editorial

Novel Processing Technology of Dairy Products

Ekaterini Moschopoulou

Laboratory of Dairy Research, Department of Food Science and Human Nutrition, Iera Odos 75, 11855 Athens, Greece; catmos@aua.gr; Tel.: +30-210-529-4680

Citation: Moschopoulou, E. Novel Processing Technology of Dairy Products. *Foods* **2021**, *10*, 2407. https://doi.org/10.3390/foods10102407

Received: 3 October 2021
Accepted: 9 October 2021
Published: 11 October 2021

Publisher's Note: MDPI stays neutral with regard to jurisdictional claims in published maps and institutional affiliations.

Milk has been processed into dairy products using traditional methods for hundreds of years.

However, research on new technologies in order to develop new products or to improve those already existing is constantly conducted. More specifically, modern processing approaches are used in order to change the texture, to improve the organoleptic properties, to ensure the safety, to extend the shelf life, and finally, to increase the nutritional and health value of dairy products. Pulsed Electric Fields, Ultrasound, and High-Pressure (HP) Processing are some of the novel technologies that are considered as alternatives to heat treatment. Among these, HP is the most promising technology during which the product is treated mainly in the range of 100–600 MPa at ambient temperatures; as a result, several constituents and properties of the treated product are altered [1]. Moreover, membrane technology is widely used in the dairy industry for separation or fractionation purposes, depending on the membrane pore size and the applied pressure. Microfiltration (MF), which involves membranes with a pore size of 0.1–10 μm, can remove bacteria and spores from skim milk; hence, it is also called 'cold pasteurization' [2]. In addition, apart from the novel technologies, modern analytical methods have been developed for the better determination or evaluation of significant characteristics and processing steps in the production of the dairy products.

The present Special Issue includes five research papers and two review papers. Four papers study the application of HP on milk and yoghurt products [3–6], one paper deals with the MF of milk [7], another one reviews the production of specific dairy ingredients [8], and the last one examines the use of modern analytical methods in the characterization of the recrystallization process in ice cream [9].

Briefly, Diez-Sánchez et al. [3] applied HP to produce milkshakes containing chokeberry pomace with the use of different high-pressure parameters and ratios of chokeberry. The researchers, with the aid of response surface methodology, concluded that a novel, highly nutritional product with increased antioxidant capacity and total phenolic content could be produced from milk that contains 10% (w/v) chokeberry and at a high pressure of 500 MPa for 10 min. The effect of HP on two probiotic microorganisms used in yoghurt production was studied by Tsevdou et al. [4]. These authors investigated different parameters and reported that HP in the range of 200–300 MPa improved the rheological properties of the product, while the viability loss of 0.5–1.2 logCFU/g remained constant during refrigeration storage. In addition, the same product enriched with a sweet cherry flavor behaved in the same way when treated with HP under the same conditions. Sakkas et al. [5] used HP as an alternative method for heating ovine milk enriched with whey protein concentrate (WPC) or whey protein concentrate hydrolysates (WPCH) in yoghurt production. WPCH had been developed by the same researchers by hydrolyzing WPC from Feta cheese whey with trypsin or Protamex. It was found that the heating method, i.e., conventional or HP, affected the yoghurt gel properties more than the enrichment with WPC or WPCH. More specifically, HP at conditions similar to heating caused inferior gel properties no matter the type of enrichment, and the addition of WPCH to skim ovine milk at a ratio of >0.5% dramatically worsened the gel properties. Wazed and Farid [6] developed a novel hypoallergenic product for infant nutrition. They applied HP to a reconstituted infant milk formula enriched with α-Lactalbumin, using different conditions of pressure, temperature, and time, and comparing it to pasteurization at 72 °C. The authors concluded that the highest ratio of α-Lactalbumin to β-Lactoglobulin level, which assure low allergenicity, was achieved after HP at 600 MPa, 40.4 °C, for 5 min.

Regarding membrane technology, Panopoulos et al. [7], who applied cross-flow MF with the use of ceramic membranes with a pore size of 1.4 μm and a transmembrane pressure of 1.5 bar at 50 °C, reported that microbial flora reduction was higher in skim ovine milk (0.4% fat) than in skim bovine milk (0.3% fat). Moreover, although ovine and bovine permeates showed lower protein content than the respective unprocessed milks, the cheesemaking properties of ovine milk were not significantly affected. The authors concluded that the application of MF on ovine milk under the studied conditions can be used to treat this milk prior to cheesemaking. Huang et al. [8], reviewing the methods for producing dairy ingredients enriched in milk phospholipids (MPLs), reported that the membrane separation process as compared to methods involving the use of organic solvents—i.e., supercritical fluid extraction (SFE) by CO_2 and dimethylether (DME), or SFE and DME, or organic solvent extraction—is the most efficient way to concentrate MPLs. Moreover, in this particular case, the carbon footprint of membrane technology is the lowest.

As far as analyses are concerned, Kamińska-Dwórznicka [9] reviewed and compared modern methods for testing and describing the recrystallization of ice crystals, a process detrimental to the quality of ice cream products. They refer to methods such as microscopy and image analysis, focused beam reflectance measurement, oscillation thermo-rheometry, nuclear magnetic resonance (NMR), splat cooling assay, and X-ray microtomography; they concluded that all presented methods are suitable for describing the recrystallization process. However, only microscopy and image analysis can show both the changes in size and shape of ice crystals as well as their location.

Finally, the editor of this Special Issue would like to thank the authors who submitted their papers and provided the readers with new information about high-pressure processing, membrane processing, and modern analytical methods for the study of the dairy products.

Conflicts of Interest: The author declares no conflict of interest.

References

1. Munir, M.; Nadeem, M.; Qureshi, T.M.; Leong, T.S.; Gamlath, C.J.; Martin, G.J.; Ashokkumar, M. Effects of high pressure, microwave and ultrasound processing on proteins and enzyme activity in dairy systems—A review. *Innov. Food Sci. Emerg. Technol.* **2019**, *57*, 102192. [CrossRef]
2. Charcosset, C. Classical and Recent Applications of Membrane Processes in the Food Industry. *Food Eng. Rev.* **2021**, *13*, 322–343. [CrossRef]
3. Diez-Sánchez, E.; Martínez, A.; Rodrigo, D.; Quiles, A.; Hernando, I. Optimizing High Pressure Processing Parameters to Produce Milkshakes Using Chokeberry Pomace. *Foods* **2020**, *9*, 405. [CrossRef]
4. Tsevdou, M.; Ouli-Rousi, M.; Soukoulis, C.; Taoukis, P. Impact of High-Pressure Process on Probiotics: Viability Kinetics and Evaluation of the Quality Characteristics of Probiotic Yoghurt. *Foods* **2020**, *9*, 360. [CrossRef] [PubMed]
5. Sakkas, L.; Tzevdou, M.; Zoidou, E.; Gkotzia, E.; Karvounis, A.; Samara, A.; Taoukis, P.; Moatsou, G. Yoghurt-Type Gels from Skim Sheep Milk Base Enriched with Whey Protein Concentrate Hydrolysates and Processed by Heating or High Hydrostatic Pressure. *Foods* **2019**, *8*, 342. [CrossRef]
6. Wazed, M.A.; Farid, M. Hypoallergenic and Low-Protein Ready-to-Feed (RTF) Infant Formula by High Pressure Pasteurization: A Novel Product. *Foods* **2019**, *8*, 408. [CrossRef] [PubMed]
7. Panopoulos, G.; Moatsou, G.; Psychogyiopoulou, C.; Moschopoulou, E. Microfiltration of Ovine and Bovine Milk: Effect on Microbial Counts and Biochemical Characteristics. *Foods* **2020**, *9*, 284. [CrossRef]
8. Huang, Z.; Zheng, H.; Brennan, C.S.; Mohan, M.S.; Stipkovits, L.; Li, L.; Kulasiri, D. Production of Milk Phospholipid-Enriched Dairy Ingredients. *Foods* **2020**, *9*, 263. [CrossRef]
9. Kamińska-Dwórznicka, A.; Gondek, E.; Łaba, S.; Jakubczyk, E.; Samborska, K. Characteristics of Instrumental Methods to Describe and Assess the Recrystallization Process in Ice Cream Systems. *Foods* **2019**, *8*, 117. [CrossRef]

Article

Optimizing High Pressure Processing Parameters to Produce Milkshakes Using Chokeberry Pomace

Elena Diez-Sánchez [1], Antonio Martínez [2], Dolores Rodrigo [2], Amparo Quiles [1] and Isabel Hernando [1,*]

[1] Department of Food Technology, Universitat Politècnica de València, Camino de Vera s/n, 46022 València, Spain; eldiesan@upvnet.upv.es (E.D.-S.); mquichu@tal.upv.es (A.Q.)
[2] Instituto de Agroquímica y Tecnología de Alimentos (IATA-CSIC), Paterna, 46980 València, Spain; amartinez@iata.csic.es (A.M.); lolesra@iata.csic.es (D.R.)
* Correspondence: mihernan@tal.upv.es

Received: 25 February 2020; Accepted: 6 March 2020; Published: 1 April 2020

Abstract: High hydrostatic pressure is a non-thermal treatment of great interest because of its importance for producing food with additional or enhanced benefits above their nutritional value. In the present study, the effect of high hydrostatic pressure processing parameters (200–500 MPa; 1–10 min) is investigated through response surface methodology (RSM) to optimize the treatment conditions, maximizing the phenol content and antioxidant capacity while minimizing microbiological survival, in milkshakes prepared with chokeberry pomace (2.5–10%). The measurement of fluorescence intensity of the samples was used as an indicator of total phenolic content and antioxidant capacity. The results showed that the intensity of the treatments had different effects on the milkshakes. The RSM described that the greatest retention of phenolic compounds and antioxidant capacity with minimum microbiological survival were found at 500 MPa for 10 min and 10% (w/v) chokeberry pomace. Therefore, this study offers the opportunity to develop microbiologically safe novel dairy products of high nutritional quality.

Keywords: antioxidant capacity; microbial inactivation; image analysis; high pressure processing; total phenolic content

---

1. Introduction

Nowadays, consumers show increasing preference for foods with additional or enhanced benefits beyond their basic nutritional value. These benefits come from the composition, e.g., bioactive compounds, which may have long-term health effects. There is convincing evidence of the cardioprotective effects for frequent consumption of fruits and vegetables high in fiber, micronutrients, and several phytochemicals. Specifically, the association between flavonoids and an increased cardiovascular health has been proven in berries [1]. Producing berry-based juice generates by-products, comprising peel and seeds, having a high nutritional value because of their polyphenol and fiber content. Berry by-products can be a value-added food ingredient [2–4] and recent studies show that the enrichment of food products with these by-products is feasible [5–7]. In this context, chokeberry (*Aronia melanocarpa*) pomace can be used, as chokeberry is one of the richest plant sources of phenolic phytochemicals, including procyanidins and anthocyanins [8,9] which are related to effectiveness in several pathological conditions where damage is caused by uncontrolled oxidative processes [10].

Previous studies have used dairy products and pomace for the production of yogurts with apple pomace [11–13] or fermented milk with grape pomace [14]. However, in the work of Issar et al. [11] and de Souza et al. [14], polyphenols and fiber were extracted, respectively, and added to milk for product preparation. Wang et al. [12,13] incorporated the apple pomace directly into the dairy matrix. Regarding berry pomace, Ni et al. [15] formulated yogurts with aqueous berry extracts from salal berry and blackcurrant pomace. In this study, we propose the incorporation of the pomace directly into the milk using high hydrostatic pressure (HPP) to help polyphenols being extracted into the matrix.

HPP is a method to preserve food and has the potential to retain or improve the bioaccessibility and bioavailability of nutritional and antioxidant compounds because of microstructural modifications [16]. HPP retains the nutritional and sensory quality of food products, but there is a concern related to food safety [17]. In this context, high pressures have been effective at inactivating vegetative cells when sufficient intense pressure is applied [18].

Thus, the present study aimed to prepare milkshakes enriched with polyphenols by adding chokeberry pomace to the milk and using HPP to improve polyphenols extraction from the pomace. The effect of high HPP parameters such as time and pressure on total phenolic content (TPC), antioxidant capacity (AC), and the microbiological inactivation in milkshakes with different concentrations of chokeberry pomace will be studied. To define the best processing conditions, response surface methodology (RSM) was used to maximize the TPC and the AC results while minimizing the microbiological survival.

2. Materials and Methods

2.1. Sample Preparation and HPP Treatments

Döhler GmbH (Darmstadt, Germany) provided fresh chokeberry pomace. The pomace was dried at 70 °C for 3 h and milled in a ZM 100 ultracentrifuge mill (Retsch GmbH, Haan, Germany) at 14,000 r.p.m. using a 0.5 mm sieve [19]. Reconstituted skimmed milk powder (Corporación Alimentaria Peñasanta S.A., Siero, Spain) was selected for chokeberry pomace inclusion.

Different concentrations of chokeberry powder were added to skimmed milk samples to give final chokeberry pomace concentrations of 2.5%, 6.25%, and 10% (*w/v*). The samples were poured into 50 mL polypropylene tubes that were introduced into polyethylene bags and heat-sealed (MULTIVAC Thermosealer, Switzerland) before undergoing HPP treatment. HPP treatments were performed in a unit with a 2.35 L vessel volume and maximum operating pressure of 600 MPa (High Pressure Food Processor, EPSI NV, Belgium). The samples were pressurized at 200, 350, and 500 MPa, at 18–22 °C, for 1, 5.5, and 10 min, using a compression rate of 300 MPa/min and a decompression time < 1 min, [20,21]. Other parameters, pressure intensity, pressurization time, and temperature were automatically controlled. Once the treatment was completed, the samples were taken from the vessel, immersed in an ice-water bath, and refrigerated (3 ± 1 °C) until use. Before each analysis, both microbiological and chemical, the samples were filtered with paper filter previously sterilized using an autoclave. The microbiological cultures and microscopic observations were immediately conducted after the filtration while the samples for the TPC and AC determination were stored by deep-freezing at −80 °C until use.

2.2. Total Phenolic Content

The total phenolic content (TPC) was determined according to the method described by Singleton et al. [22], with some modifications. The treated chokeberry pomace milk (5 mL) was homogenized in an Ultraturrax with 25 mL of 960 g kg^{-1} ethanol. The homogenate was centrifuged (4122 g, 30 min, 4 °C), filtered, and the supernatant was stored. This treatment was repeated on the leftover pellet using 25 mL of 960 g kg^{-1} ethanol to extract more supernatant, then added to the first supernatant; the total supernatant was brought up to 100 mL. After, 6 mL of distilled water and 0.5 mL of Folin-Ciocalteu reagent (1:1 (*v/v*)) were added to an aliquot of 1 mL of the ethanolic extract. After three minutes, 1 mL of sodium carbonate solution (20% (*w/v*)) (Panreac Química SLU, Castellar del Vallès, Barcelona, Spain) and 1.5 mL of distilled water were added. The mixture was vortexed and kept at room temperature (~25 °C) in a dark room for 90 min. Absorbance was measured at 765 nm using a spectrophotometer (series 1000, model CE 1021; CECIL Instruments Ltd., Cambridge, UK) with results expressed as mg of Gallic Acid Equivalents (GAE)/100 mL. Total phenolic extractions were made in triplicate.

2.3. Antioxidant Capacity

The antioxidant capacity (AC) was measured by a ferric reducing antioxidant power assay (FRAP) [23,24]. Extracts were obtained in the same way as for TPC determination. Distilled water (30 μL), the sample (30 μL), and FRAP reagent (900 μL) were placed in the cuvette. The cuvettes were incubated for 30 min in a water bath at 37 °C and the absorbance was measured at 595 nm. A calibration curve was obtained using different concentrations of Trolox in 960 g kg^{-1} ethanol. The results were expressed as μmol Trolox/mL of sample. Three separate extractions were made for each treatment and the measurements were performed in triplicate.

2.4. Chokeberry Microstructure

The microstructure analysis was carried out following Hernández-Carrión et al. [25] with some modifications. For the study of the chokeberry microstructure, a Nikon Eclipse E80i microscope (Nikon, Tokyo, Japan) was used. The autofluorescence of the phenolic compounds in the samples was observed using a mercury arc lamp with a tetramethyl rhodamine filter (λ_{ex} = 543/22 nm, λ_{em} = 593/40 nm) as the excitation source. Samples were visualized using ×10 and ×20 objective lenses. The micrographs were stored at a 1280 × 1024-pixel resolution using the microscope software (NIS-Elements F, Version 4.0, Nikon, Tokyo, Japan). Analysis of the fluorescence intensity was conducted with the ImageJ software.

2.5. Microbiological Analysis

The effect of HPP treatment was evaluated on natural contaminating flora (aerobic mesophilic microorganisms, molds, and yeasts) and on *Listeria monocytogenes* serovar 4b [26] (CECT 4032) as pathogen microorganism [27]. The growth media used for the spreading of samples was plate count agar (Scharlau Chemie S. A., Sentmenat, Spain) for mesophilic aerobic; potato dextrose agar (Scharlau Chemie S. A., Sentmenat, Spain) for molds and yeasts; and tryptic soy agar (Scharlau Chemie S. A., Sentmenat, Spain) for *L. monocytogenes*. The incubation conditions were 48 h at 30 °C, 120 h (5 days) at 24 °C, and 48 h at 37 °C, respectively.

L. monocytogenes was artificially inoculated in the sample. The stock vials containing *L. monocytogenes* at a concentration ca. 10^9 cfu/mL were generated following the methods described by Saucedo-Reyes et al. and Pina-Pérez et al. [20,28]. Before HPP treatment, vials were inoculated in the chokeberry-skimmed milk samples at a final concentration of 10^8 cfu/mL. The counts for evaluating microorganism inactivation were performed before and after each HPP treatment. Two aliquots (0.1 mL) were taken from each sample, diluted with buffered peptone water (Scharlau Chemie S. A, Sentmenat, Spain), and plated in the respective agar. Two replicas of each treatment were obtained, and three repetitions of each treatment condition was performed. The survival fraction S = N/N$_0$ and the level of inactivation Log$_{10}$ (N/N$_0$) were evaluated for each repetition.

2.6. Experimental Design and Statistical Analysis

RSM was used to optimize the preservation process and to investigate the simultaneous effects of pressure, time, and chokeberry powder concentration on TPC, AC, and microbiological inactivation of the prepared samples. For the chokeberry-milk matrix, a face-centered central composite design was used with three levels (maximum, minimum, and central) and three independent factors, namely pressure (200 to 500 MPa), time (1 to 10 min), and chokeberry pomace concentration (2.5 to 10% (*w/v*)), resulting in 16 combinations (Table 1). The central point of the three variables was replicated twice to assure the reproducibility and stability of the results. All the experiments were randomized. A quadratic model was obtained with regression coefficients associated with the linear, quadratic, and interaction effects. A *t*-test determined their significance through the *p*-value generated.

Table 1. Experimental design matrix for studies conducted.

Run	Pressure (MPa) (X_1)	Time (min) (X_2)	Chokeberry Pomace (% (*w/v*))
1	500	1	2.5
2	350	10	6.25
3	500	5.5	6.25
4 [a]	350	5.5	6.25
5	500	10	10
6	350	5.5	2.5
7	350	5.5	10
8	350	1	6.25
9	200	10	10
10	500	10	2.5
11	200	5.5	6.25
12	200	10	2.5
13	350	5.5	6.25
14	200	1	10
15	200	1	2.5
16	500	1	10

[a] Central point.

The non-significant terms ($p > 0.05$) were deleted from the second-order polynomial model after an ANOVA test and a new ANOVA was performed to find the coefficients of the final equation for better accuracy [29]. The experimental design and the data analysis were performed using the Statgraphics® Centurion XVII software (Statpoint Technologies, Inc., Warrenton, VA, USA).

3. Results and Discussion

3.1. Effect of HPP on TPC, AC, and Microbial Counts of Chokeberry Milkshakes

Effects of HPP treatments on the TPC, AC, and microbiological survival fraction are shown in Table 2. The analyses were conducted on untreated and treated samples to observe differences with HPP treatment. Results for molds and yeasts are not shown because there were no changes in any treatments.

TPC concentration in untreated samples with 2.5%, 6.25%, and 10% (*w/v*) of chokeberry pomace is 53.02 ± 0.14, 73.92 ± 3.17, and 121.16 ± 0.31 mg GAE/100 mL, respectively. Furthermore, the AC results for samples 2.5, 6.25, and 10% (*w/v*) of chokeberry pomace, are 6.06 ± 0.14, 9.27 ± 0.20, and 14.89 ± 0.30 µmol Trolox/mL, respectively. As expected, the TPC and AC results are higher with higher pomace concentrations.

In treated samples, the highest result for TPC is 155.28 ± 2.07 mg GAE/100 mL with 500 MPa during 1 min and 10% (*w/v*) pomace addition; the lowest is 42.45 ± 2.89 mg GAE/100 mL with 350 MPa during 5.5 min and 2.5% (*w/v*) pomace addition. Yet, the highest AC is 17.3 ± 1.08 µmol Trolox/mL for the treatment of 200 MPa during 1 min with 10% (*w/v*) pomace addition. The treatment that obtained the lower AC matches the one that obtained the lower TPC (350 MPa during 5.5 min with 2.5% (*w/v*) pomace addition).

The lowest results for TPC and AC are obtained for the intermediate pressure and not for the lowest as expected, yet these low results are similar to other treatments at different processing conditions but with the same pomace concentration (2.5%). In contrast, the higher value results for TPC and AC are obtained for milkshakes with 10% of chokeberry. When comparing the results of the treatments at each concentration, values were higher as the pomace concentration increased. In addition, comparing the results of treatments at each concentration with its untreated counterpart, samples 6.25 and 10% show an increase in TPC and AC, influenced by the pressures and times studied. However, for samples at 2.5%, this effect is less acute, affected only by high pressures (500 MPa).

Table 2. Effect of HPP and chokeberry pomace on TPC, AC, and the microbial survival fraction.

Pressure (MPa)	Time (min)	Chokeberry Pomace % (w/v)	TPC (mg GAE/100 mL)	AC (μmol Trolox/mL)	Inactivation Log_{10} (N/N$_0$)
0	0	2.5	53.02 ± 0.14	6.06 ± 0.14	-
		6.25	73.92 ± 3.17	9.27 ± 0.20	-
		10	121.16 ± 0.31	14.89 ± 0.30	-
200	1	2.5	49.39 ± 2.24	5.02 ± 0.34	0.01 ± 0.08
		10	130.20 ± 3.46	17.3 ± 1.08	−0.18 ± 0.04
	5.5	6.25	136.79 ± 8.06	11.79 ± 0.99	−0.20 ± 0.08
	10	2.5	50.54 ± 4.64	4.80 ± 0.16	0.06 ± 0.12
		10	132.85 ± 2.31	13.98 ± 0.49	−0.11 ± 0.06
350	1	6.25	84.32 ± 2.59	10.13 ± 0.79	−0.20 ± 0.06
	5.5	2.5	42.45 ± 2.89	4.77 ± 0.39	−0.25 ± 0.14
		6.25 [a]	78.54 ± 3.39 [a]	7.58 ± 0.41 [a]	−0.21 ± 0.19 [a]
		6.25	106.14 ± 4.28	8.54 ± 0.03	−0.33 ± 0.09
		10	124.96 ± 3.78	16.45 ± 0.10	−0.55 ± 0.05
	10	6.25	101.92 ± 5.70	16.53 ± 0.78	−0.33 ± 0.32
500	1	2.5	54.14 ± 0.61	5.7 ± 0.45	−0.44 ± 0.05
		10	155.28 ± 2.07	16.36 ± 0.35	−0.67 ± 0.08
	5.5	6.25	97.17 ± 7.09	11.33 ± 2.03	−0.87 ± 0.06
	10	2.5	58.36 ± 2.97	6.69 ± 0.19	−3.63 ± 0.08
		10	134.17 ± 1.57	14.79 ± 0.76	−4.02 ± 0.15

[a] Central point; HPP: High Pressure Processing; TPC: Total Phenolic Content; AC: Antioxidant Capacity; N: final cell concentration; N$_0$: initial cell concentration.

Thus, the results are affected by all the factors in this study (pressure, time, and concentration), but mainly the pomace concentration. The HPP treated foods are either unaffected or have increased TPC and/or extractability following treatments with high pressures [30]. Andrés et al. [31] found increases of 6.6% and 4.2% in TPC values for fruit smoothies treated at 450 and 600 MPa, respectively. Corrales et al. [32] showed that treating at 600 MPa enhanced the anthocyanin extraction and its AC in grape by-products than with conventional extraction methods. Liu et al. [33] found treatments at 200 MPa, for 5 and 10 min, led to an increased TPC of 14.24% and 14.35% in wild *Lonicera caerulea* berry, respectively, however, for treatments at 500 and 600 MPa there was a significant decrease of TPC. In contrast, other authors found HPP had little effect on phenolic content. Barba et al. [34] observed TPC to be relatively resistant to the effect of processing in tomato purées. Hurtado et al. [35] did not observe differences in AC values between untreated and treated red fruit-based smoothies for treatments at 350 MPa, 10 °C, and 5 min. Patras et al. [36] found that phenol content in HPP treated strawberry purées was relatively resistant to the effect of processing at 400 and 500 MPa, only showing an increase in treatments at 600 MPa for TPC and AC. Therefore, the results obtained with HPP depend of several conditions, such as the matrix in which they are applied, and the severity of the treatment and it is necessary to study the behavior of different samples with these treatments.

In fluorescence microscopy, the intensity value of a pixel is related to the number of fluorophores present at the corresponding area in the particle. Thus, the digital images can be used to extract the intensity values to determine the local concentration of fluorophores in a specimen [37]. In our case of study, the images in Figure 1 show the pomace particles dispersed into the milk matrix.

To analyze the florescence intensity the images corresponding to the lower (200 MPa, 1 min, 2.5%), central (350 MPa, 5.5 min, 6.25%), and higher (500 MPa, 10 min, 10%) treatments were selected. The particle with greater intensity was selected to generate intensity profiles (Figure 1a–c). A line (shown in yellow) was drawn across the particle, and a plot (graph) was generated to show the intensity values of the pixels along the line (Figure 1d–i). In addition, Figure 1j–l shows the relation between the percentages of particles at each fluorescence intensity interval.

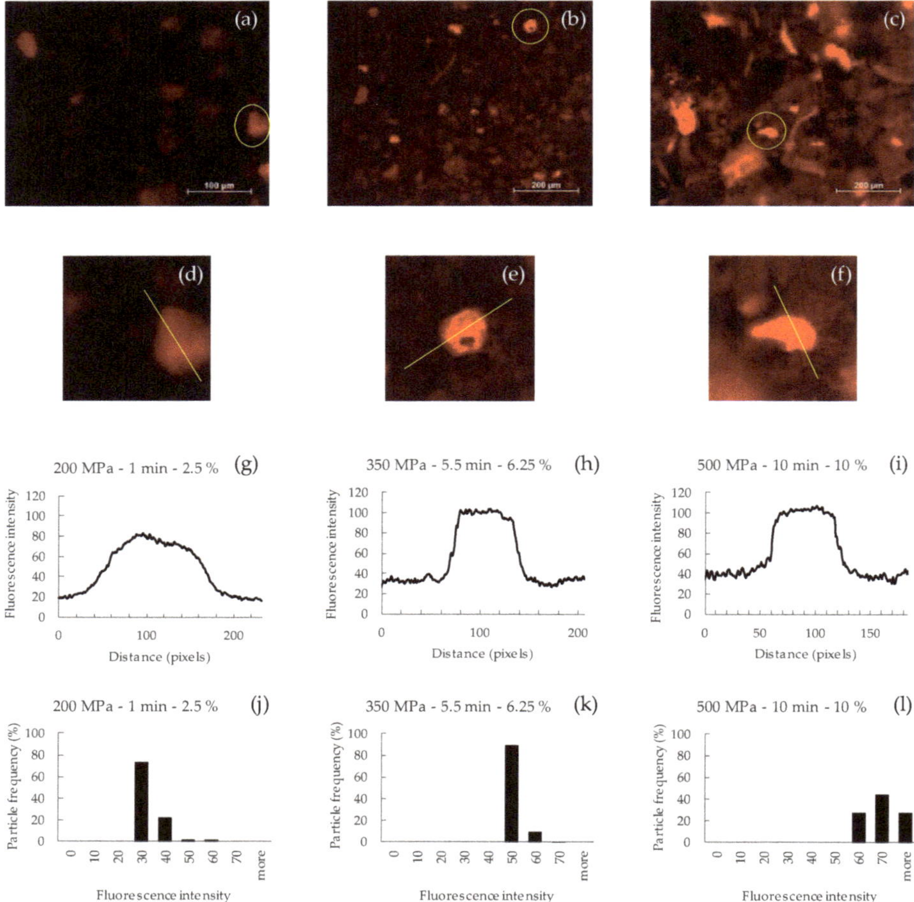

Figure 1. Fluorescence intensity of: 200 MPa, 1 min, 2.5% (**a,d,g,j**); 350 MPa, 5.5 min, 6.25% (**b,e,h,k**); and 500 MPa, 10 min, 10% (**c,f,i,l**).

The fluorescence intensity for the isolated particles is higher in the medium (Figure 1k) and high treatments (Figure 1l) than in the low treatment (Figure 1j). Comparing the background intensity, corresponding to the liquid phase of the sample, fluorescence increases as the severity of the treatment increases. Several authors [16,38,39] have described the cell wall degradation and breakage in plant tissue after HPP, leading to a leaching of contents from the pomace cells (included polyphenols) to the milk acting as a liquid medium [32,38]. In addition, Gonzalez and Barrett [40] described that treatments at pressures above 220 MPa were responsible for the breakage of the membrane structure because of protein unfolding and interface separation. Therefore, as higher pressures are applied, the phenolic contents are being released to the medium due to the membrane breakage, giving as a result higher values of fluorescence intensity. The particle frequency plots show that the frequency of particles at high intensities rises with the severity of the treatment. These results agree with the results of TPC and AC, i.e., higher fluorescence intensities correspond to the treatments that obtained the higher TPC and AC results. Therefore, measurement of fluorescence intensity can be an indicator for TPC and AC in this type of sample. Further research is needed to prove if the analysis is usable in other matrices.

Besides the effect of HPP on the polyphenols, there could be a microbial inactivation because of changes induced in the microbial cells. These changes include alterations in the cell membrane, effects on proteins, and effects on the genetic mechanism of microorganisms [18,41,42]. Seen in microbiological inactivation results in Table 2, treatments at 200 MPa during 1 min with 2.5% (*w/v*) of pomace and at 200 MPa during 10 min with 2.5% of pomace do not show microbial inactivation. At 2.5% pomace concentration and low pressure (200 MPa), longer treatment time is not enough for microbial inactivation. Muñoz-Cuevas et al. [43] also observed this behavior. Still, it is necessary to reach a minimum treatment intensity (500 MPa, 10 min) to obtain significant *L. monocytogenes* inactivation. At this condition, an increase in chokeberry pomace concentration from 2.5% to 10% (*w/v*) increases microbial inactivation from 3.63 to 4.02 Log reductions.

Thus, increasing the pressure and treatment time results in an increase in the lethal effect of HPP treatment. This effect relates to food composition, technological parameters, and the factors acting in synergy [44,45].

Besides the effect of treatment conditions, several authors have described the high antimicrobial capacity of berry fruits and their by-products [46–48] and the synergistic effect between natural substances and high pressure treatments [28,49,50]. Despite the evidence found in the literature, except the treatments of 500 MPa, 10 min with 2.5% and 10% (*w/v*) pomace, the inactivation values are lower than similar treatments with other products and microorganisms. For example, Evrendilek & Balasubramaniam [49] concluded that samples of ayran (yogurt drink) treated at 600 MPa during 5 min had reduced in the levels of *L. monocytogenes* and *L. innocua* by more than five log units ($p < 0.05$) at ambient temperature. Nevertheless, Gervilla et al. [51] and Black et al. [52] have described a baroprotective effect that milk has on the cells. Thus, this effect could counteract the antimicrobial effect of chokeberry pomace, explaining the low inactivation levels found for *L. monocytogenes* in this study. To prove this effect, an experiment was conducted where the central point of the design (350 MPa, 5.5 min in milk with 6.25% (*w/v*) of chokeberry pomace) was used as a treatment to compare the inactivation reached in four different matrices: (i) peptone water with *L. monocytogenes*, (ii) milk with *L. monocytogenes*, (iii) peptone water with *L. monocytogenes* and chokeberry pomace, and (iv) milk with *L. monocytogenes* and chokeberry pomace. Results tested the baroprotective effect of milk and are shown in Figure 2.

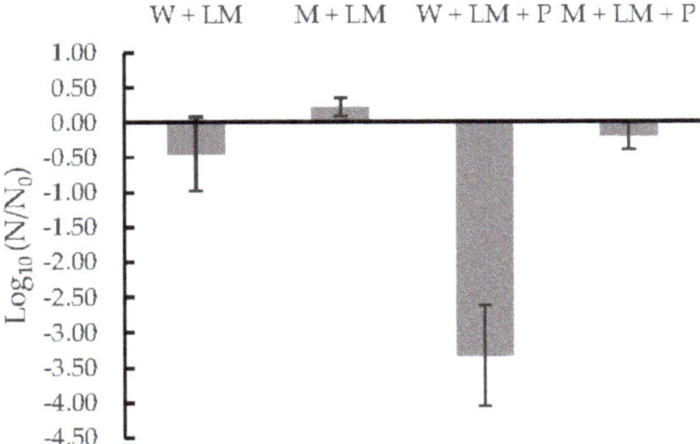

Figure 2. Inactivation level of ingredients' different combinations: peptone water (W), milk (M), inoculated *Listeria monocytogenes* (LM), and pomace (P).

In samples without milk (W + LM and W + LM + P), the number of surviving cells is lower than with milk samples (M + LM and M + LM + P), and more prominent when pomace is added. Apart from the protective effect of milk, an increase is seen in the efficacy of HPP against *L. monocytogenes* when pomace is present in the sample. Thus, these results can describe the synergistic effect of pomace,

HPP, and the protective effect of milk on *L. monocytogenes*. Still, there is microbial inactivation with some treatments, even with the protective effect of milk on the microbiological cells.

3.2. Processing Parameter Optimization and Their Effect on the Safety and Quality of the Formulated Matrix

The best processing conditions for treating chokeberry milkshakes when HPP is combined with added by-products with antimicrobial and antioxidant properties (chokeberry pomace) were studied by RSM. This methodology uses a sequence of designed experiments to obtain an optimal response.

First, the estimated effects of each factor (pressure, time, and concentration) and their interactions were analyzed (Figure 3). The response function for the factors and the adjusted regression coefficient (R^2 adjusted), showing the percentage of variation in the response explained by the fitted model, is shown in Equations (1)–(3) (pressure (P), time (t), chokeberry pomace concentration (C)). The value of the adjusted R^2 close to one indicates a high correlation between the experimental and fitted values.

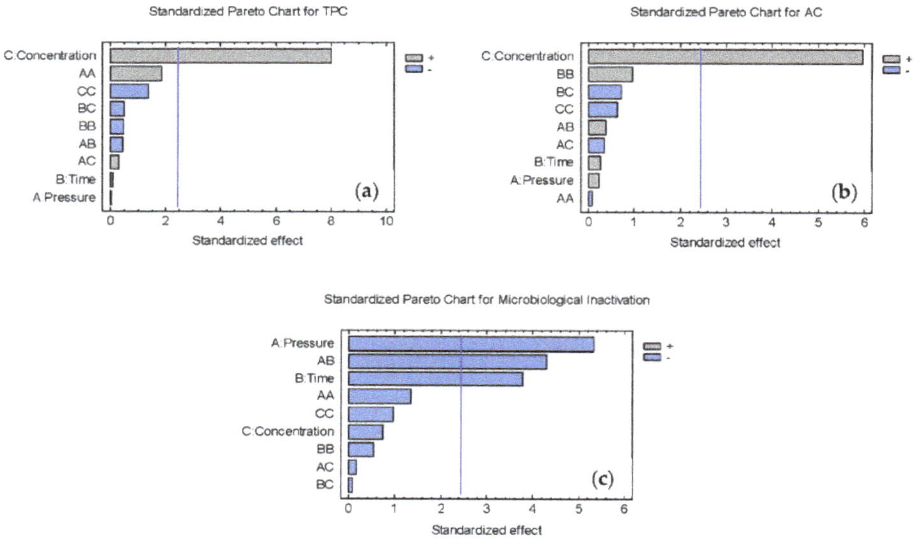

Figure 3. Estimated effects of each factor (pressure, time, and concentration) and their interactions with TPC (**a**), AC (**b**), and *L. monocytogenes* inactivation (**c**) where + or − means a positive or negative relation between factor (pressure, time or concentration) and response (TPC, AC or *L. monocytogenes* survival fraction), respectively. A: pressure, B: time and C: concentration. The combination of letters (AA, BB, AB …) refer to the interactions carried out in the analysis.

$$\text{TPC} = 25.6475 + 11.2687 \times \text{C} \qquad R^2\text{adj} = 0.85 \qquad (1)$$

$$\text{AC} = 2.08477 + 1.38395 \times \text{C} \qquad R^2\text{adj} = 0.80 \qquad (2)$$

$$\text{Log}_{10}\left(\frac{\text{N}}{\text{N}_0}\right) = -0.160277 + 0.000618685 \times \text{P} + 0.285165 \times \text{t} - 0.00123019 \times \text{P} \times \text{t} \quad R^2\text{adj} = 0.76 \quad (3)$$

Figure 3 shows the pareto chart for TPC, AC, and microbial inactivation. This chart determines the magnitude and the importance of the effects. The bars that extend beyond the line correspond to effects that are statistically significant with a 95.0% confidence level. The factor "pomace concentration in milkshakes" is the only factor that significantly affects TPC and AC concentration. However, the results in Table 2 show TPC and AC are influenced by all the factors, including time and pressure. Chokeberry pomace has been reported as a berry fruit with high phenolic content [8]. Thus, though it could exist with the effect of pressure and time, the results could be masked by the natural high phenolic content.

The low effect of treatment conditions on TPC and AC could be explained by using milk as a liquid medium. High pressure processing induces physicochemical and technological changes in milk properties. When HPP is applied to milk, the casein micelles are disintegrated into casein particles of smaller size, which is accompanied by an increase in casein and calcium phosphate levels in the serum phase of milk and by a decrease in both non-casein nitrogen and serum nitrogen fractions [18,53]. In addition, interactions between polyphenols and milk proteins have been previously described by other researchers [54–56]. In our work, these interactions could be favored by the changes in the casein structure due to HPP treatment, which would lead to the formation of complexes that restrict the accessibility of analysis, leading to lower AC and TPC and a non-significant effect of treatment conditions (pressures and time). Tadapaneni et al. [57] also observed this effect in strawberry-based beverages treated with HPP at pressures ranging from 200 to 600 MPa. They saw, when formulated with milk instead of water, the beverage presented reduced levels of AC and anthocyanins because of complexes forming between polyphenols and milk proteins. Therefore, as the effect of concentration is so pronounced in RSM and polyphenol–milk protein interactions may exist, decreasing the AC and TPC, the effect of the other parameters is much lower, leading to a non-significant effect of time and pressure.

In contrast, pressure and time are the parameters with a significant effect on the microbiological inactivation. Thus, the chokeberry pomace concentration added to the milkshake, does not have a significant effect on the inactivation results. These results confirm the hypothesis explained above (Figure 2); there is an antimicrobial effect of berry pomace. However, it is masked with the protective effect of milk on *L. monocytogenes* cells, giving as a lower inactivation result than with similar treatment conditions in products with a natural antimicrobial agents, yet without milk [49]. Despite the protective effect of milk on microorganisms, adding chokeberry pomace could help achieve higher inactivation levels than HPP without the pomace.

Once the estimated effect and its interaction were analyzed, the response optimization was carried out. The results show that the optimized factors are 500 MPa for 10 min in milk with 10% (*w/v*) of chokeberry pomace (Table 3). This treatment condition ensures the maximum TPC and AC with the minimum microbiological survival.

Table 3. Predicted and limit response values for optimum treatment conditions.

Response	Predicted	Lower 95.0% Limit	Upper 95.0% Limit
TPC (mg GAE/100 mL)	138.33	125.68	150.99
AC (μmol Trolox/mL)	15.92	14.13	17.72
Log_{10} (N/N$_0$)	−3.15	−3.96	−2.34

The optimum treatment conditions are the same as those of the experimental design. When comparing the results in Table 2 with the predicted values in Table 3, we see that experimental results are like the predicted values through optimization. Therefore, the RSM is proven to be a reliable tool to predict the behavior of the sample studied in terms of AC, TPC, and microbial inactivation.

4. Conclusions

The fluorescence intensity measurement of microstructure images can be an indicator of the TPC and AC of the samples. Microstructure images showed that, with intense treatment, there is leaching of the polyphenolic substances into the milk because of cell structure breakage. For the microbiological inactivation, the results showed that the pomace had antimicrobial properties, but they were partially masked by the interactions between milk proteins and the polyphenols available, and that higher levels of inactivation were achieved at high pressures and long treatment times. The RSM results showed that TPC and AC were only affected by the pomace concentration added to the milkshake, because the high pomace concentration and the polyphenol–milk protein interactions could mask the effect of pressure and time.

Although the efficiency of HPP inactivation on *L. monocytogenes* has been proven, further research is needed for products without milk to study the effect of chokeberry pomace treated with HPP on the TPC, AC, and microbial inactivation.

Author Contributions: Formal analysis, D.R.; Investigation, E.D.-S.; Methodology, A.M.; Supervision, A.Q.; Writing–review & editing, I.H. All authors have read and agreed to the published version of the manuscript.

Funding: This research was funded by Instituto Nacional de Investigación y Tecnología Agraria y Alimentaria (INIA-Spain) through the "BERRYPOM. Adding value to processing waste: innovative ways to incorporate fibers from berry pomace in baked and extruded cereal-based foods" project included in the ERA-NET-SUSFOOD program, to Spanish Ministry of Science, Innovation and Universities through project AGL2017-86840-C2-2-R, and to FEDER funds from EU.

Acknowledgments: The authors are grateful to Instituto Nacional de Investigación y Tecnología Agraria y Alimentaria (INIA-Spain) for financial support through the BERRYPOM. Adding value to processing waste: innovative ways to incorporate fibers from berry pomace in baked and extruded cereal-based foods project included in the ERA-NET-SUSFOOD program, to Spanish Ministry of Science, Innovation and Universities through project AGL2017-86840-C2-2-R, and to FEDER funds from EU. They would also like to thank Phillip John Bentley for his assistance in correcting the manuscript's English.

Conflicts of Interest: The authors declare no conflict of interest.

References

1. Basu, A.; Rhone, M.; Lyons, T.J. Berries: Emerging impact on cardiovascular health. *Nutr. Rev.* **2010**, *68*, 168–177. [CrossRef]
2. Tańska, M.; Roszkowska, B.; Czaplicki, S.; Borowska, E.J.; Bojarska, J.; Dąbrowska, A. Effect of fruit pomace addition on shortbread cookies to improve their physical and nutritional values. *Plant Foods Hum. Nutr.* **2016**, *71*, 307–313. [CrossRef] [PubMed]
3. Rodríguez-Werner, M.; Winterhalter, P.; Esatbeyoglu, T. Phenolic composition, radical scavenging activity and an approach for authentication of *Aronia melanocarpa* berries, juice, and pomace. *J. Food Sci.* **2019**, *84*, 1791–1798. [CrossRef] [PubMed]
4. De Souza, D.R.; Willems, J.L.; Low, N.H. Phenolic composition and antioxidant activities of saskatoon berry fruit and pomace. *Food Chem.* **2019**, *290*, 168–177. [CrossRef] [PubMed]
5. Diez-Sánchez, E.; Quiles, A.; Llorca, E.; Reißner, A.-M.; Struck, S.; Rohm, H.; Hernando, I. Extruded flour as techno-functional ingredient in muffins with berry pomace. *LWT* **2019**, *113*, 108300. [CrossRef]
6. Quiles, A.; Llorca, E.; Schmidt, C.; Reißner, A.-M.; Struck, S.; Rohm, H.; Hernando, I. Use of berry pomace to replace flour, fat or sugar in cakes. *Int. J. Food Sci. Technol.* **2018**, *53*, 1579–1587. [CrossRef]
7. Ospina, M.; Montaña-Oviedo, K.; Díaz-Duque, Á.; Toloza-Daza, H.; Narváez-Cuenca, C.-E. Utilization of fruit pomace, overripe fruit, and bush pruning residues from Andes berry (*Rubus glaucus* Benth) as antioxidants in an oil in water emulsion. *Food Chem.* **2019**, *281*, 114–123. [CrossRef]
8. Kulling, S.; Rawel, H. Chokeberry (*Aronia melanocarpa*)—A review on the characteristic components and potential health effects. *Planta Med.* **2008**, *74*, 1625–1634. [CrossRef]
9. Kähkönen, M.P.; Hopia, A.I.; Heinonen, M. Berry phenolics and their antioxidant activity. *J. Agric. Food Chem.* **2001**, *49*, 4076–4082. [CrossRef]
10. Chrubasik, C.; Li, G.; Chrubasik, S. The clinical effectiveness of chokeberry: A systematic review. *Phyther. Res.* **2010**, *24*, 1107–1114. [CrossRef]
11. Issar, K.; Sharma, P.C.; Gupta, A. Utilization of apple pomace in the preparation of fiber-enriched acidophilus yoghurt. *J. Food Process. Preserv.* **2017**, *41*, e13098. [CrossRef]
12. Wang, X.; Kristo, E.; LaPointe, G. Adding apple pomace as a functional ingredient in stirred-type yogurt and yogurt drinks. *Food Hydrocoll.* **2020**, *100*, 105453. [CrossRef]

13. Wang, X.; Kristo, E.; LaPointe, G. The effect of apple pomace on the texture, rheology and microstructure of set type yogurt. *Food Hydrocoll.* **2019**, *91*, 83–91. [CrossRef]

14. de Souza de Azevedo, P.O.; Aliakbarian, B.; Casazza, A.A.; LeBlanc, J.G.; Perego, P.; de Souza Oliveira, R.P. Production of fermented skim milk supplemented with different grape pomace extracts: Effect on viability and acidification performance of probiotic cultures. *PharmaNutrition* **2018**, *6*, 64–68. [CrossRef]

15. Ni, H.; Hayes, H.E.; Stead, D.; Raikos, V. Incorporating salal berry (*Gaultheria shallon*) and blackcurrant (*Ribes nigrum*) pomace in yogurt for the development of a beverage with antidiabetic properties. *Heliyon* **2018**, *4*, e00875. [CrossRef]

16. Vázquez-Gutiérrez, J.L.; Plaza, L.; Hernando, I.; Sánchez-Moreno, C.; Quiles, A.; de Ancos, B.; Cano, M.P. Changes in the structure and antioxidant properties of onions by high pressure treatment. *Food Funct.* **2013**, *4*, 586. [CrossRef]

17. San Martín, M.F.; Barbosa-Cánovas, G.V.; Swanson, B.G. Food processing by high hydrostatic pressure. *Crit. Rev. Food Sci. Nutr.* **2002**, *42*, 627–645. [CrossRef]

18. Welti-Chanes, J.; López-Malo, A.; Palou, E.; Bermúdez, D.; Guerrero-Beltrán, J.A.; Barbosa-Cánovas, G.V. Fundamentals and applications of high pressure processing to foods. In *Novel Food Processing Technologies*; Barbosa-Cánovas, G.V., Tapia, M.S., Cano, M.P., Martín-Belloso, O., Martínez, A., Eds.; CRC Press: Boca Raton, FL, USA, 2005; pp. 164–170. ISBN 0-8257-5333-X.

19. Reißner, A.M.; Al-Hamimi, S.; Quiles, A.; Schmidt, C.; Struck, S.; Hernando, I.; Turner, C.; Rohm, H. Composition and physicochemical properties of dried berry pomace. *J. Sci. Food Agric.* **2019**, *99*, 1284–1293. [CrossRef]

20. Saucedo-Reyes, D.; Marco-Celdrán, A.; Pina-Pérez, M.C.; Rodrigo, D.; Martínez-López, A. Modeling survival of high hydrostatic pressure treated stationary- and exponential-phase *Listeria innocua* cells. *Innov. Food Sci. Emerg. Technol.* **2009**, *10*, 135–141. [CrossRef]

21. Pina Pérez, M.C.; Rodrigo Aliaga, D.; Saucedo Reyes, D.; Martínez López, A. Pressure inactivation kinetics of *Enterobacter sakazakii* in infant formula milk. *J. Food Prot.* **2007**, *70*, 2281–2289. [CrossRef]

22. Singleton, V.L.; Orthofer, R.; Lamuela-Raventós, R.M. Analysis of total phenols and other oxidation substrates and antioxidants by means of folin-ciocalteu reagent. *Methods Enzymol.* **1999**, *299*, 152–178.

23. Benzie, I.F.F.; Strain, J.J. The ferric reducing ability of plasma (FRAP) as a measure of "antioxidant power": The FRAP assay. *Anal. Biochem.* **1996**, *239*, 70–76. [CrossRef] [PubMed]

24. Pulido, R.; Bravo, L.; Saura-Calixto, F. Antioxidant activity of dietary polyphenols as determined by a modified ferric reducing/antioxidant power assay. *J. Agric. Food Chem.* **2000**, *48*, 3396–3402. [CrossRef] [PubMed]

25. Hernández-Carrión, M.; Sanz, T.; Hernando, I.; Llorca, E.; Fiszman, S.M.; Quiles, A. New formulations of functional white sauces enriched with red sweet pepper: A rheological, microstructural and sensory study. *Eur. Food Res. Technol.* **2015**, *240*, 1187–1202. [CrossRef]

26. Murray, E.G.D.; Webb, R.A.; Swann, M.B.R. A disease of rabbits characterised by a large mononuclear leucocytosis, caused by a hitherto undescribed bacillus *Bacterium monocytogenes* (n.sp.). *J. Pathol. Bacteriol.* **1926**, *29*, 407–439. [CrossRef]

27. Farber, J.M.; Peterkin, P.I. *Listeria monocytogenes*, a food-borne pathogen. *Microbiol. Rev.* **1991**, *55*, 476–511. [CrossRef]

28. Pina-Pérez, M.C.; Silva-Angulo, A.B.; Muguerza-Marquínez, B.; Aliaga, D.R.; López, A.M. Synergistic effect of high hydrostatic pressure and natural antimicrobials on inactivation kinetics of *bacillus cereus* in a liquid whole egg and skim milk mixed beverage. *Foodborne Pathog. Dis.* **2009**, *6*, 649–656. [CrossRef]

29. Barba, F.J.; Criado, M.N.; Belda-Galbis, C.M.; Esteve, M.J.; Rodrigo, D. *Stevia rebaudiana* Bertoni as a natural antioxidant/antimicrobial for high pressure processed fruit extract: Processing parameter optimization. *Food Chem.* **2014**, *148*, 261–267. [CrossRef]

30. Tokuşoğlu, Ö. Effect of high hydrostatic pressure processing strategies on retention of antioxidant phenolic bioactives in foods and beverages—A review. *Polish J. Food Nutr. Sci.* **2016**, *66*, 243–251. [CrossRef]

31. Andrés, V.; Villanueva, M.J.; Tenorio, M.D. The effect of high-pressure processing on colour, bioactive compounds, and antioxidant activity in smoothies during refrigerated storage. *Food Chem.* **2016**, *192*, 328–335. [CrossRef]

32. Corrales, M.; Toepfl, S.; Butz, P.; Knorr, D.; Tauscher, B. Extraction of anthocyanins from grape by-products assisted by ultrasonics, high hydrostatic pressure or pulsed electric fields: A comparison. *Innov. Food Sci. Emerg. Technol.* **2008**, *9*, 85–91. [CrossRef]

33. Liu, S.; Xu, Q.; Li, X.; Wang, Y.; Zhu, J.; Ning, C.; Chang, X.; Meng, X. Effects of high hydrostatic pressure on physicochemical properties, enzymes activity, and antioxidant capacities of anthocyanins extracts of wild *Lonicera caerulea* berry. *Innov. Food Sci. Emerg. Technol.* **2016**, *36*, 48–58. [CrossRef]

34. Barba, F.J.; Esteve, M.J.; Frigola, A. Ascorbic acid is the only bioactive that is better preserved by high hydrostatic pressure than by thermal treatment of a vegetable beverage. *J. Agric. Food Chem.* **2010**, *58*, 10070–10075. [CrossRef] [PubMed]

35. Hurtado, A.; Guàrdia, M.D.; Picouet, P.; Jofré, A.; Ros, J.M.; Bañón, S. Stabilisation of red fruit-based smoothies by high-pressure processing. Part II: Effects on sensory quality and selected nutrients. *J. Sci. Food Agric.* **2017**, *97*, 777–783. [CrossRef]

36. Patras, A.; Brunton, N.P.; Da Pieve, S.; Butler, F. Impact of high pressure processing on total antioxidant activity, phenolic, ascorbic acid, anthocyanin content and colour of strawberry and blackberry purées. *Innov. Food Sci. Emerg. Technol.* **2009**, *10*, 308–313. [CrossRef]

37. Waters, J.C. Accuracy and precision in quantitative fluorescence microscopy. *J. Cell Biol.* **2009**, *185*, 1135–1148. [CrossRef]

38. Gao, G.; Ren, P.; Cao, X.; Yan, B.; Liao, X.; Sun, Z.; Wang, Y. Comparing quality changes of cupped strawberry treated by high hydrostatic pressure and thermal processing during storage. *Food Bioprod. Process.* **2016**, *100*, 221–229. [CrossRef]

39. Hernández-Carrión, M.; Hernando, I.; Quiles, A. High hydrostatic pressure treatment as an alternative to pasteurization to maintain bioactive compound content and texture in red sweet pepper. *Innov. Food Sci. Emerg. Technol.* **2014**, *26*, 76–85. [CrossRef]

40. Gonzalez, M.E.; Barrett, D.M. Thermal, high pressure, and electric field processing effects on plant cell membrane integrity and relevance to fruit and vegetable quality. *J. Food Sci.* **2010**, *75*, R121–R130. [CrossRef]

41. Patterson, M.F. Microbiology of pressure-treated foods. *J. Appl. Microbiol.* **2005**, *98*, 1400–1409. [CrossRef]

42. Ritz, M.; Tholozan, J.; Federighi, M.; Pilet, M. Physiological damages of *Listeria monocytogenes* treated by high hydrostatic pressure. *Int. J. Food Microbiol.* **2002**, *79*, 47–53. [CrossRef]

43. Muñoz-Cuevas, M.; Guevara, L.; Aznar, A.; Martínez, A.; Periago, P.M.; Fernández, P.S. Characterisation of the resistance and the growth variability of Listeria monocytogenes after high hydrostatic pressure treatments. *Food Control* **2013**, *29*, 409–415. [CrossRef]

44. Ferreira, M.; Almeida, A.; Delgadillo, I.; Saraiva, J.; Cunha, Â. Susceptibility of *Listeria monocytogenes* to high pressure processing: A review. *Food Rev. Int.* **2016**, *32*, 377–399. [CrossRef]

45. Possas, A.; Pérez-Rodríguez, F.; Valero, A.; García-Gimeno, R.M. Modelling the inactivation of *Listeria monocytogenes* by high hydrostatic pressure processing in foods: A review. *Trends Food Sci. Technol.* **2017**, *70*, 45–55. [CrossRef]

46. Bartkiene, E.; Lele, V.; Sakiene, V.; Zavistanaviciute, P.; Ruzauskas, M.; Bernatoniene, J.; Jakstas, V.; Viskelis, P.; Zadeike, D.; Juodeikiene, G. Improvement of the antimicrobial activity of lactic acid bacteria in combination with berries/fruits and dairy industry by-products. *J. Sci. Food Agric.* **2019**, *99*, 3992–4002. [CrossRef]

47. Cisowska, A.; Wojnicz, D.; Hendrich, A.B. Anthocyanins as antimicrobial agents of natural plant origin. *Nat. Prod. Commun.* **2011**, *6*, 149–156. [CrossRef]

48. Nohynek, L.J.; Alakomi, H.-L.; Kähkönen, M.P.; Heinonen, M.; Helander, I.M.; Oksman-Caldentey, K.-M.; Puupponen-Pimiä, R.H. Berry phenolics: Antimicrobial properties and mechanisms of action against severe human pathogens. *Nutr. Cancer* **2006**, *54*, 18–32. [CrossRef]

49. Evrendilek, G.A.; Balasubramaniam, V.M. Inactivation of *Listeria monocytogenes* and *Listeria innocua* in yogurt drink applying combination of high pressure processing and mint essential oils. *Food Control* **2011**, *22*, 1435–1441. [CrossRef]

50. Vurma, M.; Chung, Y.-K.; Shellhammer, T.H.; Turek, E.J.; Yousef, A.E. Use of phenolic compounds for sensitizing *Listeria monocytogenes* to high-pressure processing. *Int. J. Food Microbiol.* **2006**, *106*, 263–269. [CrossRef]

51. Gervilla, R.; Capellas, M.; Ferragut, V.; Guamis, B. Effect of high hydrostatic pressure on *Listeria innocua* 910 CECT inoculated into ewe's milk. *J. Food Prot.* **1997**, *60*, 33–37. [CrossRef]

52. Black, E.P.; Huppertz, T.; Fitzgerald, G.F.; Kelly, A.L. Baroprotection of vegetative bacteria by milk constituents: A study of *Listeria innocua*. *Int. Dairy J.* **2007**, *17*, 104–110. [CrossRef]

53. Bordenave, N.; Hamaker, B.R.; Ferruzzi, M.G. Nature and consequences of non-covalent interactions between flavonoids and macronutrients in foods. *Food Funct.* **2014**, *5*, 18–34. [CrossRef] [PubMed]

54. Zhang, H.; Yu, D.; Sun, J.; Liu, X.; Jiang, L.; Guo, H.; Ren, F. Interaction of plant phenols with food macronutrients: Characterisation and nutritional–physiological consequences. *Nutr. Res. Rev.* **2014**, *27*, 1–15. [CrossRef] [PubMed]

55. Jakobek, L. Interactions of polyphenols with carbohydrates, lipids and proteins. *Food Chem.* **2015**, *175*, 556–567. [CrossRef] [PubMed]

56. Tadapaneni, R.K.; Banaszewski, K.; Patazca, E.; Edirisinghe, I.; Cappozzo, J.; Jackson, L.; Burton-Freeman, B. Effect of high-pressure processing and milk on the anthocyanin composition and antioxidant capacity of strawberry-based beverages. *J. Agric. Food Chem.* **2012**, *60*, 5795–5802. [CrossRef] [PubMed]

57. Chawla, R.; Patil, G.R.; Singh, A.K. High hydrostatic pressure technology in dairy processing: A review. *J. Food Sci. Technol.* **2011**, *48*, 260–268. [CrossRef]

 *foods*

Article

Impact of High-Pressure Process on Probiotics: Viability Kinetics and Evaluation of the Quality Characteristics of Probiotic Yoghurt

Maria Tsevdou [1], Maria Ouli-Rousi [1], Christos Soukoulis [2] and Petros Taoukis [1,*]

[1] Laboratory of Food Chemistry and Technology, School of Chemical Engineering, National Technical University of Athens, 5 Heroon Polytechniou Str., 157 80 Athens, Greece; mtsevdou@chemeng.ntua.gr (M.T.); maria.ouli@gmail.com (M.O.-R.)
[2] Luxembourg Institute of Science and Technology, Systems and Bioprocessing Engineering Group, Environmental Research and Innovation, 5 avenue des Hauts Fourneaux, L4362 Esch-sur-Alzette, Luxembourg; christos.soukoulis@list.lu
* Correspondence: taoukis@chemeng.ntua.gr; Tel.: +30-210-772-3171

Received: 14 February 2020; Accepted: 17 March 2020; Published: 19 March 2020

Abstract: The impact of high-pressure (HP) processing on the viability of two probiotic microorganisms (*Bifidobacterium bifidum* and *Lactobacillus casei*) at varying pressure (100–400 MPa), temperature (20–40 °C) and pH (6.5 vs. 4.8) conditions was investigated. Appropriate mathematical models were developed to describe the kinetics of the probiotics viability loss under the implemented HP conditions, aiming to the development of a predictive tool used in the design of HP-processed yoghurt-like dairy products. The validation of these models was conducted in plain and sweet cherry-flavored probiotic dairy beverage products pressurized at 100–400M Pa at ambient temperature for 10 min. The microbiological, rheological, physicochemical and sensory characteristics of the HP-treated probiotic dairy beverages were determined in two-week time intervals and for an overall 28 days of storage. Results showed that the application of HP in the range of 200–300 MPa had minimal impact on the probiotic strains viability throughout the entire storage period. In addition, the aforementioned HP processing conditions enhanced the rheological and sensory properties without affecting post-acidification compared to the untreated product analogues.

Keywords: probiotics; viability model; high-pressure processing; rheology; sensory quality; fermented dairy beverage

---

1. Introduction

In recent decades, the functional food market has experienced a remarkable expansion as a result of the increasing consumer awareness for products that confer significant well-being and health promoting benefits. According to Food and Agriculture Organization/World Health Organization (FAO/WHO) definition, the term probiotics refers to "live microorganisms, which when administered in adequate amounts confer a health benefit to the host" [1]. In this context, only well-defined commensals and microbe consortia isolated from human samples with generic or core effects on gut physiology and supporting the health of reproductive tract, oral cavity, lungs, skin or brain-gut axis can be considered to be probiotics [2]. As for functional food innovation, fermented products such as yoghurt, cheese, fermented vegetables, fruit and legumes and dry cured meat are considered to be indigenous sources of probiotics [3]. The health benefits attained by the regular consumption of probiotic foods are associated with postbiotics, i.e., the production of secondary metabolites such as organic acids, enzymes, bioactive or antimicrobial peptides, exopolysaccharides, conjugated linoleic acids, vitamins, and phenolic compounds. Current knowledge of probiotics supports a plethora of therapeutic effects achieved following their regular administration to the human host, including their ability to relieve the symptomatology of irritable bowel syndrome, improve the blood serum lipid composition, stimulate the gut immunomodulation and prevent inflammation induced chronic disease such as obesity and several forms of cancer [4].

As far as the food industry practices, preserving the biological activity of probiotics is quite challenging as several endogenous, i.e., food matrix associated, and exogenous parameters, such as exposure of bacterial cells to harsh food processing, storage, and post-ingestion conditions, can potentially act as sublethal stressors of probiotic cells. For example, the availability of nutrients, bacterial cell growth promoters or inhibitors, the physical state (i.e., rubbery or glassy), the amount of dissolved oxygen, the pH and water activity conditions are among the commonest food matrix associated parameters that affect the viability of probiotics. In addition, food processing and storage can result in significant mechanical, heat, pH, osmolytic and pro-oxidants exposure induced injuries of the bacterial cells [5].

Despite their well-addressed biofunctional profile, fermented dairy products need to meet a handful of quality characteristics relating to texture, structure, olfactory, gustatory, and visual sensory modalities. In this context, the addition of food relevant structuring and texturizing agents, such as skim-milk powder, whey protein concentrates (WPCs), caseinates, polysaccharides, or other bulking agents, is considered to be a standard practice in dairy products manufacturing [6,7]. As an alternative, high-pressure (HP) processing has shown a great potential to deliver bespoke texture and structure reinforcing benefits to protein gels including fermented milk products [8–13]. It has been demonstrated that the implementation of a HP processing in a pressure-temperature range from 100 to 400 MPa and from 10 to 25 °C, respectively, and for an overall processing time of about 10–15 min, is sufficient to promote fermented milks structuring without the need for solid fortification. In addition, the aforementioned HP processing conditions can impart acceptable sensory cross-modality without affecting in an adverse way the biological activity of the living yoghurt starters and probiotic cells [14–16].

In the present work, it is hypothesized that the HP processing of pre-coagulated milk (stirred yoghurt) can lead to beneficial effects on its major physicochemical, rheological, and sensory properties without affecting significantly the viability of the embedding probiotic cells. For this purpose, the effect of HP processing conditions on the viability of two commonly used probiotic microorganisms (*Bifidobacterium bifidum* and *Lactobacillus casei*) in different pH values (pH 6.5 and 4.8) model systems was kinetically studied. The developed predictive models were assessed for their feasibility in the design of HP-processed real food systems, i.e., plain or cherry fruit-flavored probiotic yoghurts, containing *Bifidobacterium lactis* BB12 and *Lactobacillus acidophilus* LA5 strains. The probiotic yoghurts were evaluated concerning the microbiological, physicochemical, rheological, and sensory characteristics for an overall of 28 days of storage at chilling conditions.

2. Materials and Methods

2.1. Preparation of the Inocula for Kinetics Study

Stock cultures of the tested microorganisms (*Bifidobacterium bifidum* and *Lactobacillus casei*) were maintained in cryovials at −40 °C with glycerol (20% *v/v*) used as a cryoprotective agent. For the revival of each microorganism, one cryovial was transferred into 10 mL MRS broth (Merck, 1.10661, Darmstadt, Germany) and incubated at 37 °C for 24 h; for the revival of *Bifidobacterium bifidum*, 3% *v/v* of L-cysteine HCl sol. was added to MRS broth to achieve anaerobic environment. For growth and use in the kinetic experiments, 100 µL of the above inocula were transferred individually into 10 mL MRS broth (with 3% *v/v* of L-cysteine HCl sol. for *Bifidobacterium bifidum*) and incubated at 37 °C for 18–20 h. The final suspensions were transferred into 90 mL MRS broth with modified pH value to 4.80 (HCl sol. 1 M) or 6.50 (Na_2HPO_4/NaH_2PO_4 buffer sol. 0.1 M), representing the model system of low and high pH value respectively, and served as the inocula for viability loss experiments (microbial cells in stationary phase and initial plate counted at approximately 10^9 CFU/mL). The selection of MRS growth medium instead of a milk-based medium was based on the fact that the viability of the tested microorganisms will not be affected by the possible presence of any protective agent (e.g., lipids) of the bacterial cells.

2.2. Yoghurt Preparation

Commercial homogenized pasteurized milk (3.2% *w/w* protein and 3.5% *w/w* fat) was placed into 2000 mL glass beakers and subjected to thermal treatment consisting of a rapid microwave assisted pre-heating step at 85 °C and a batch heating step at 85 °C for 20 min in a water-bath (Memmert, Guechenbach, Germany). Subsequently, milk was cooled to fermentation temperature (43.0±0.2 °C) and inoculated with 0.2% *v/v* commercial starter culture comprising *Streptococcus salivarius* subsp. *thermophilus* and *Lactobacillus delbrueckii* subsp. *bulgaricus* in a ratio of 4:1 (prepared as a 1:5 *w/w* dilution of Christian–Hansen freeze-dried YC-X11 culture in commercial UHT skim milk) and 0.3% *w/w* commercial probiotic cultures of *Bifidobacterium lactis* and *Lactobacillus acidophilus* (direct inoculation of Christian–Hansen BB12 and LA-5 probiotic cultures) according to the supplier's instructions. The inoculated milk was placed in a water-bath maintained at the desirable fermentation temperature (43.0±0.4 °C) until the acidification end point (pH = 4.80) was achieved. Afterwards, the coagulum was manually stirred and divided in three batches. Two of the batches were single stage homogenized at 10 bar in a laboratory high-pressure homogenizer (APV1000, Kolding, Denmark) and packed in 180 g sterile multilayer pouches (laminate film: PP-aluminum-PE); one batch was stored at 5.0 ± 0.2 °C (**Control samples**), while the other was subjected to HP treatment (homogenized and HP-treated samples, coded as **Homo-HP**) and then stored under the same chilling conditions. A third batch prepared by directly packing the yoghurt in sterile multilayer pouches of 180 g for HP experiments (HP-treated samples, coded as **HP**) was also prepared. For cherry-flavored samples, 10% (*w/w*) of cherry syrup (Olympic Foods S.A., Athens, Greece) was blended with the yoghurt base until homogeneous dispersion of the syrup (68.6±1.57% *w/v* total sugars content of the syrup). The day after the production of the samples was considered to be the zero time point, while all quality characterization analysis were carried out in triplicate on a week time intervals for an overall of 28 days.

2.3. High-Pressure Processing

The experiments were performed using a pilot-scale HP equipment (Food Pressure Unit FPU 1.01, Resato International BV, Assen, Netherlands), as previously described in Tsevdou et al. [12]. For kinetic experiments, 5 mL of the inocula were placed into pouches (laminate film: PP-aluminum-PE) and the viability loss experiments were conducted in the multi-vessel system, in duplicate at various combinations of pressure (100−400 MPa) and temperature (20−40 °C) for appropriate process times. For product batch experiments, 180 g of yoghurt were placed into multilayer pouches (laminate film: PP-aluminum-PE) and, the experiments were conducted in triplicate at various combinations of pressure (100−400 MPa) and ambient temperature (26.1±0.3 °C) for 10 min.

2.4. Microbiological Analysis

Ten grams of yoghurt was transferred into a sterile stomacher bag with 90 g sterilized Ringer solution (1.15525, Merck, Darmstadt, Germany) and was homogenized for 60 s with a stomacher (BagMixer ® Interscience, Saint-Nom-la-Bretèche, France). The enumeration of remaining viable cells of 10-fold serial dilutions of yoghurt homogenates was conducted using the appropriate plate methodology.

For enumerating the probiotic bacteria, MRS agar (Merck, 1.10660, Darmstadt, Germany) was modified as follows; (a) addition of 5% *v/v* filter-sterilized NNLP (NNLP sol. comprises 100 mg Neomycin sulfate, 15 mg Nalidixic acid, 3 g Lithium chloride and 200 mg Paromomycin sulfate per 1 L (Tharmaraj & Shah, 2004)) sol. and 3% *v/v* filter-sterilized L-cysteine HCl sol. for Bifidobacteria spp., and (b) addition of 0.2% *v/v* filter-sterilized vancomycin hydrochloride sol. (Calbiochem, 627850, Darmstadt, Germany) for *Lactobacillus casei* enumeration [17]. Incubation of petri dishes was carried out in anaerobic jars (Merck, 1.16387, Darmstadt, Germany) at 37 °C for 72 h, using Anaerocult A (Merck, 1.13829, Darmstadt, Germany) as a catalyst. *Streptococcus thermophilus* was enumerated on M17 Agar (1.15108, Merck, Darmstadt, Germany) after incubation at 37 °C for 24 h under aerobic conditions. For *Lactobacillus bulgaricus* enumeration, the pour plate methodology on MRS agar with modified pH value at 4.58 was used, followed by incubation at 45 °C for 72 h in anaerobic jars as previously described. Two replicates of at least three appropriate dilutions were enumerated.

2.5. Physicochemical Analysis

The pH of yoghurt samples was measured using a pH meter (AMEL 338, Amel Instruments, Milano, Italy) whereas their titratable acidity (expressed as % lactic acid) was determined according to the International Dairy Federation (IDF) method [18].

The water holding capacity (WHC) of yoghurts was expressed as the grams of separated whey from 10 g of sample after centrifugation at 10.000 rpm (5.712 g) and 20 °C for 20 min [19].

The total color (E) of the samples were determined by measurement of CIELab values (L-value: lightness, a-value: redness/greenness, b-value: yellowness/blueness), using a CR200-Minolta Chromameter (Minolta Co., Chuo-Ku, Osaka, Japan) with an 8 mm measuring area, according to the equation:

$$E = \sqrt{L^2 + a^2 + b^2} \tag{1}$$

The instrument was standardized using a white reference tile (Minolta Co., Chuo-Ku, Osaka, Japan). All measurements were carried out in triplicate.

2.6. Rheological Properties

The rheological behavior of yoghurt samples was determined using a rotational viscosimeter (RC1 Rheometer, Rheotec Meßtechnic GmbH, Raderburg, Germany) coupled with a circulating cooling bath (RE312, Lauda GmbH, Lauda-Königshofen, Germany). The measurements were carried out at 10 °C using a MS-CC48 DIN/FTK cylinder. 75 mL of yoghurt sample were transferred to the measuring cup and preconditioned at 10 °C for 1 h prior to analysis. Shear rate sweeps from 5 to 200 s^{-1} followed by constant shear rate step at 200 s^{-1} were applied. The duration of the shear stress sweep was 180 s. To describe the rheological behavior of the samples, the shear stress – shear rate data were fitted to the Ostwald–de Waale model:

$$\sigma = K \cdot \gamma^n \tag{2}$$

where, σ is the shear stress (Pa·s), γ is the shear rate (s^{-1}), K is the consistency index (Pa·s^n) and n is the flow behavior index. The apparent viscosity (η, Pa·s) of the samples was calculated at a shear rate of 50 s^{-1}, representing the sensing shear rate in the mouth of low viscosity foods [20].

2.7. Sensory Evaluation

Eight trained panelists (according to ISO standards) belonging to the staff of the Laboratory of Food Chemistry and Technology were recruited for assessing the plain and cherry-flavored yoghurts on days 1, 15 and 28 [21,22].

Two sessions per day (one in the morning and one in the afternoon) evaluating five formulations per session were conducted in the accredited according to ISO 17.025 Sensory Laboratory of NTUA (Athens, Greece) that has a standardized room (according to ISO standards) equipped with separate booths [23]. Samples were presented in plastic coded (three-digit random codes) cups, containing 50 mL of freshly removed from the refrigerator yoghurt samples. The panelists were asked to evaluate the samples using a 9-point intensity scale (1, lowest intensity; 9, highest intensity) based on pre-selected appearance (wheying-off, white color, yellowish color, cherry fruit color), tactile (ropy, uniform coagulum), orotactile (viscous, curdy, grainy), gustatory (sweet, bitter, sour, metallic) and olfactory (dairy flavor, sourmilk flavor, cherry flavor, rancid) sensory modalities. In addition, the panelists were asked to rate the yoghurt samples using a 9-point hedonic scale (1: I do not like at all, 9: I like extremely). During the sensory evaluation sessions, the panelists were instructed to cleanse their palate with low sodium spring water (Zagori, Greece) and consume a small piece of unsalted bread. The data were collected in specifically designed ballots and the panelists were encouraged to write down any criticisms on the tested products.

2.8. Data Analysis

First order kinetics was fitted to the logarithm of the concentration of viable cells [24]. The decimal reduction times (D, min) were estimated to describe the effect of process time on the viability loss of the tested microorganisms. The effect of temperature was expressed through the thermal resistance constant (z_T), and the effect of pressure was described by the pressure resistance constant (z_P) [24]. The parameters of the proposed mathematical model, which describes the viability loss of the microorganisms as a function of pressure and temperature, were estimated using non-linear regression on SYSTAT 10.0 software (SYSTAT 10.0 Statistics, 2002, SPSS Inc., Chicago, USA).

2.9. Statistical Analysis

The normal distribution of the data was verified by means of the Shapiro–Wilk test and Q-Q plot representation. In addition, the equality of variance within among the variables was verified using the Levene's test. 3-factors ANOVA (treatment vs. pressure level vs. storage time) followed by Duncan's means post hoc comparison test was applied for the analysis of all quality attributes in the shelf-life study of yoghurt samples. The physicochemical, rheological and sensory data matrices were log transformed and subject to Principal Components Analysis (PCA) to explore the qualitative affinities between the different products developed. All statistical analyses were performed using Statistica® v.7 software (StatSoft Inc., Tulsa, OK, USA). Partial least squares regression (PLSR) using the leave-one-out validation method, as previously described in Kanta et al. [25], was used in order to explore the sensory modal drivers of degree of overall liking (DOL) of yoghurt samples. All PLS analyses were carried out using XLSTAT 2014 software (Addinsoft, UK).

3. Results and Discussion

3.1. Probiotics Viability Loss as a Function of Pressure and Temperature

The viability loss of *Bifidobacterium bifidum* as a function of pressurization time in model systems of different pH values at various combinations of pressure (100–400 MPa) and temperature (20–40 °C) was described by first order kinetics (R^2 0.84–0.99). The D-values were estimated at all studied pressure/temperature combinations (Table 1). The D-values decreased with increasing processing pressure and temperature at all levels tested, indicating the combined effect of temperature and pressure on the viability loss of these bacteria. In addition, *B. bifidum* exerted better survivability rates in high pH conditioned model systems, which corroborates previous studies reporting an optimal pH growth threshold in the range of 6.0–7.0 and a significant loss of survival for pH values below 4.5–5.0 [26,27].

At each temperature, the effect of pressure on the viability loss of *B. bifidum* was expressed through the pressure resistance constant (z_P) (Table 1). When conducting the experiments in acid model system (4.80), the z_P value ranged from 84.7 MPa for processing at 20 °C to 98 MPa for processing at 35 °C (R^2 range: 0.88–0.97), while in model system with pH value close to the optimum pH of growth (6.50), the z_P values ranged from 119 MPa when processing at 25 °C to 152 MPa when processing at 40 °C (R^2 range: 0.89–0.99). Taking into account that in both laboratory and industrial scale HP equipment the minimum pressure setting step is 50 MPa, the z_P value for each model system was considered to be constant for the temperature range.

Table 1. Viability loss kinetic parameters of *Bifidobacterium bifidum*.

	Decimal Reduction Times (D, min) of *B. bifidum* in pH 4.80				
	20 °C	**25 °C**	**30 °C**	**35 °C**	**z_T (°C)**
100 MPa	435 (±23)	400 (± 19)	263 (± 5.2)	250 (± 7.3)	**62.2 (± 16)**
200 MPa	313 (±9.0)	40.3 (± 1.4)	18.6 (± 2.0)	16.2 (± 0.7)	**11.9 (± 3.2)**
300 MPa	4.08 (±0.01)	2.53 (± 0.04)	1.43 (± 0.1)	0.71 (± 0.06)	**19.8 (± 1.7)**
400 MPa	0.25 (±0.03)	0.27 (± 0.01)	0.24 (± 0.01)	0.11 (± 0.00)	**37.5 (± 3.4)**
z_P (MPa)			**90.9 (± 5.1)**		
	Decimal Reduction Times (D, min) of *B. bifidum* in pH 6.50				
	25 °C	**30 °C**	**35 °C**	**40 °C**	**z_T (°C)**
100 MPa	-	526 (± 15)	333 (± 2.1)	185 (± 9.0)	22.1 (± 2.3)
200 MPa	303 (± 3.0)	128 (± 10)	108 (± 9.3)	95.2 (± 7.0)	78.4 (± 9.4)
300 MPa	38.6 (± 2.1)	36.5 (± 1.9)	35.2 (± 0.6)	15.8 (± 2.0)	39.7 (± 1.6)
400 MPa	6.47 (± 1.1)	6.20 (± 0.9)	2.22 (± 0.3)	2.09 (± 0.4)	28.9 (± 1.2)
z_P (MPa)			**151 (± 8.0)**		

Values are means ± standard error of regression.

At each pressure, the effect of temperature on the viability loss of *B. bifidum* on the D-values was estimated using the Arrhenius equation and expressed in bacteriological terms by the thermal resistance constant (z_T) (Table 1). All z_T values were estimated for all pressures tested (R^2 range: 0.81–0.99), and it was observed that there is no specific trend when increasing the applied pressure, which is in accordance with findings in the literature for a variety of microorganisms [24,28].

The viability loss of *Lactobacillus casei* was studied in growth media of both pH values. Results indicate that in growth medium with pH value of 6.50 (Figure 1a) *L. casei* exhibited similar viability loss behavior to that of *B. bifidum*, regardless the pressure applied. However, when *L. casei* was studied in the acid growth medium (Figure 1b), significant difference ($p < 0.05$) was observed in its viability as compared to that of *B. bifidum*, especially in the pressure range of 100−200 MPa. When high pressures were applied (300−400 MPa) both probiotic microorganisms exhibited similar viability loss. Similar results were obtained for all the pressure/temperatures combinations tested.

The z_P value of *L. casei* was found similar of that of *B. bifidum* when tested in acid environment (91.7 ± 1.74 MPa in the temperature range of 20–35 °C, R^2 range: 0.92–0.94), while when the experiments conducted in model system of high pH value (6.50), *L. casei* found to be slightly, yet significantly ($p < 0.01$), more baroresistant than the *B. bifidum* is (z_P value equal to 208 ± 6.53 MPa in the temperature range of 25−40 °C (R^2 range: 0.95–0.96). There are no available data in the literature that could explain the observed baroresistance of Lactobacilli; however, several shelf-life studies in fermented dairy products have showed that both microorganisms lose their viability in pH values below 4.5−5.0, and that in higher pH values Lactobacilli are significantly less sensitive as compared to Bifidobacteria [29–31]. All z_T values of Lactobacilli were estimated for all pressures tested and, as in the case of *B. bifidum*, no specific trend was observed while increasing the applied pressure, with z_T values ranging from 15.8 to 45.1 °C and from 23.4 to 40.7 °C when the bacteria were studied in growth medium of pH 4.80 and 6.50, respectively.

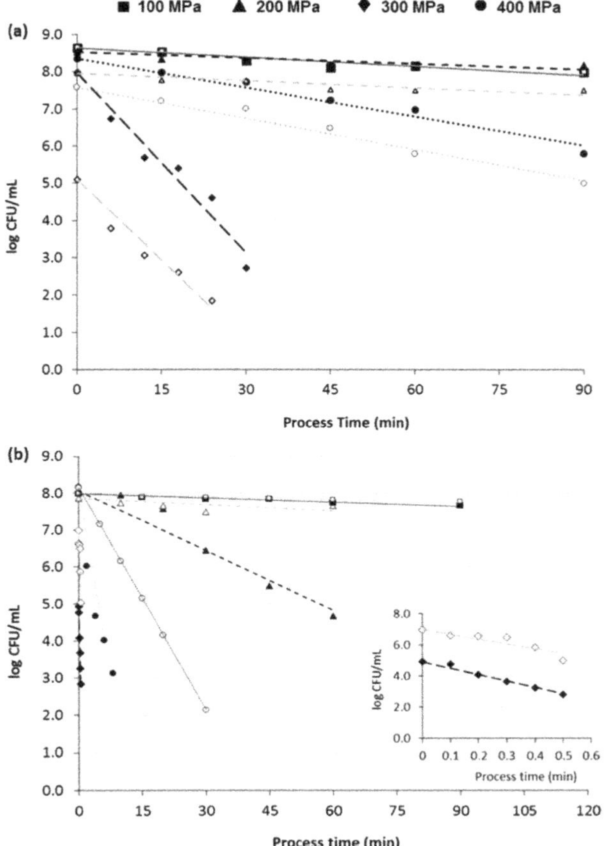

Figure 1. Comparison of *B. bifidum* (closed symbols-solid lines) and *L. casei* (open symbols-dashed lines) viability loss in broth model systems of pH value of (**a**) 6.5 and (**b**) 4.8, when pressurized at 30 °C (mean values ± standard deviation; linear lines represent the first order kinetics that describes the viability loss as a function of process time).

3.2. Modeling Probiotics Viability Loss as a Function of Temperature and Pressure

A single multi-parameter model was developed to describe the effect of pressure and temperature process conditions only on the D-value of Bifidobacteria;

$$D = D_0 \cdot \left(\exp\left\{ -\frac{2.303 \cdot T \cdot T_{ref}}{z_T} \cdot \exp\left[-A\left(P - P_{ref}\right)\right] \cdot \left(\frac{1}{T} - \frac{1}{T_{ref}}\right) + \frac{2.303}{z_P} \cdot \left(P - P_{ref}\right) \right\} \right)^{-1} \qquad (3)$$

where D_0 (min) is the decimal reduction time at the reference conditions of pressure (P_{ref}, MPa) and temperature (T_{ref}, °C), A is a constant parameter of the proposed model (MPa^{-1}) and z_T (°C) and z_P (MPa) are the thermal and pressure resistance constants, respectively.

This equation takes into account the effect of pressure on the z_T value, while the z_P value found not to be dependent on the process temperature. The parameters of the model were estimated (Table 2) using non-linear regression and, results indicated that the predicted D-values from the model were well correlated with the corresponding D-values obtained from the experimental data (Figure 2, R^2 0.97–0.99).

Table 2. Viability loss kinetic parameters of *Bifidobacterium bifidum*.

Parameter	Estimated Value pH 4.80	Estimated Value pH 6.50
P_{ref} (MPa)	200	200
T_{ref} (°C)	25	25
D_0 (min)	44.5 ± 6.39	281 ± 19.7
z_T (°C)	5.91 ± 0.44	22.9 ± 3.50
z_P (MPa)	90 (constant)	140 (constant)
A (MPa^{-1})	0.016 ± 0.002	0.002 ± 0.000
R^2	0.99	0.95

Values are means ± standard error of regression.

As detailed described above, considering that the survivability of *B. bifidum* was more sensitive to pressure and temperature conditions contrary to this of *L. casei*, we presumably deduce that the specific model can be a useful tool for estimating the survival of single or symbiotic bacterial cultures, comprising strains of the Bifidobacteria and Lactobacilli species, when exposed to HP processing conditions.

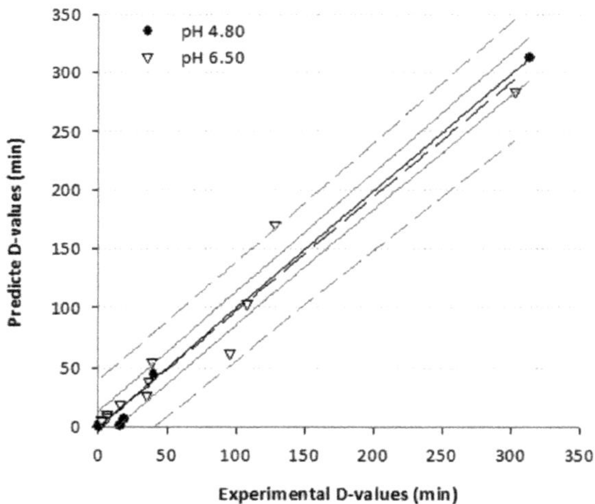

Figure 2. Correlation of experimental and predicted from the proposed model decimal reduction times (D) of *B. bifidum* tested in growth media of different pH values (Solid and dashed black lines represent the linear correlation of the data at pH of 4.8 and 6.5 respectively, while grey solid and dashed lines represent the corresponding 95% prediction bands).

3.3. Selection of Optimal HP Conditions-Application in Yoghurt Production

Given that the application of HP processing at the pressure range of 100–300 MPa and temperature range of 20–25 °C for 10–15 min did not induce any significant lethality to the probiotic cells population under acidic conditions, we decided to assess the feasibility of these processing conditions when applied to a fermented probiotic dairy food. Treatment of yoghurt samples was also performed at 400 MPa, to explore possible positive effect on the rheological parameters at this higher pressure. The physicochemical, rheological and sensory properties as well as the probiotics cells survivability over a 28 days storage period were monitored.

3.3.1. Viability of Starter Culture and Probiotic Bacteria in Yoghurt Samples

The total viable counts of the starter culture (*Str. thermophilus* and *L. bulgaricus*) in the plain probiotic yoghurt samples after HP treatment and during storage are given in Table 3.

Table 3. Viability of starter culture (as total numbers of *Str. thermophilus* and *L. bulgaricus*) in different treated plain and cherry-flavored yoghurt during storage.

Plain Yoghurt				
Control *	-			
D + 1	8.9 ± 0.2 [i]			
D + 15	8.8 ± 0.1 [hi]	-		-
D + 28	8.7 ± 0.0 [gh]			
HP	**100 MPa**	**200 MPa**	**300 MPa**	**400 MPa**
D + 1	8.7 ± 0.1 [fgh]	8.7 ± 0.1 [fgh]	8.4 ± 0.2 [bcd]	N.D.
D + 15	8.5 ± 0.1 [defg]	8.2 ± 0.3 [abc]	8.5 ± 0.1 [def]	N.D.
D + 28	8.5 ± 0.1 [def]	8.2 ± 0.1 [ab]	8.5 ± 0.1 [def]	N.D.
Homo-HP	**100 MPa**	**200 MPa**	**300 MPa**	**400 MPa**
D + 1	8.7 ± 0.2 [gh]	8.6 ± 0.1 [efgh]	8.6 ± 0.0 [defgh]	N.D.
D + 15	8.5 ± 0.0 [defgh]	8.1 ± 0.1 [a]	8.6 ± 0.0 [defgh]	N.D.
D + 28	8.4 ± 0.0 [cde]	8.1 ± 0.1 [a]	8.5 ± 0.1 [defg]	N.D.
Cherry-Flavored Yoghurt	-	-	-	-
Control *	-	-	-	-
D + 1	8.8 ± 0.0 [ij]			
D + 15	8.4 ± 0.0 [bcd]	-	-	-
D + 28	8.3 ± 0.1 [bc]			
HP	**100 MPa**	**200 MPa**	**300 MPa**	**400 MPa**
D + 1	8.9 ± 0.0 [j]	8.6 ± 0.0 [fgh]	8.5 ± 0.1 [cde]	N.D.
D + 15	8.8 ± 0.1 [ij]	8.2 ± 0.1 [a]	8.3 ± 0.1 [b]	N.D.
D + 28	8.5 ± 0.0 [def]	8.3 ± 0.1 [bc]	8.5 ± 0.1 [cde]	N.D.
Homo-HP	**100 MPa**	**200 MPa**	**300 MPa**	**400 MPa**
D + 1	8.7 ± 0.1 [hi]	8.7 ± 0.1 [hi]	8.6 ± 0.1 [fg]	N.D.
D + 15	8.6 ± 0.0 [gh]	8.5 ± 0.0 [cde]	8.4 ± 0.0 [bc]	N.D.
D + 28	8.5 ± 0.1 [efg]	8.3 ± 0.0 [b]	8.2 ± 0.0 [a]	N.D.

* Control samples were not HP-treated but only homogenized at 10 bar after the break of the coagulum. Letter D indicates the day of production. N.D. indicates microbial load below the acceptable limit of 7.0 $\log_{10}$ CFU/g of the total microbial counts for the starter culture at the end of shelf life. Different letters indicate significant difference ($p < 0.05$) between tested samples according to Duncan's mean values post hoc comparison test.

HP processing of the probiotic yoghurts induced a reduction of the starter culture total load accounting for ca. 0.2–0.5 $\log_{10}$ CFU/g for HP-treated products and ca. 0.2–0.3 $\log_{10}$CFU/g for Homo-HP-treated products. Throughout storage, starter culture cells underwent a pressure dependent decrease in their viability of about 0.2–0.5 $\log_{10}$CFU/g; yet their total counts were well above the recommended level of 7.0 $\log_{10}$ CFU/g [32]. When probiotic yoghurts were high pressurized at 400 MPa, (in the absence or presence of conventional homogenization step), a greater decrease in starter culture total counts was observed which were well below the WHO/FAO acceptable thresholds.

The remaining probiotic counts of probiotic dairy beverages after HP treatment and during storage are shown in Table 4 for *Bifidobacterium lactis* BB12 and in Figure 3 for *Lactobacillus acidophilus* LA5.

Table 4. Viability of *Bifidobacterium lactis* BB12 in different treated plain and cherry-flavored yoghurt during storage.

Plain Yoghurt				
Control *	-		-	
D + 1	8.2 ± 0.1 [jk]			
D + 15	8.3 ± 0.2 [jk]	-		-
D + 28	8.3 ± 0.1 [k]			
HP	**100 MPa**	**200 MPa**	**300 MPa**	**400 MPa**
D + 1	8.1 ± 0.1 [hijk]	7.9 ± 0.0 [fghi]	6.8 ± 0.3 [c]	N.D.
D + 15	8.1 ± 0.1 [ghijk]	7.8 ± 0.1 [fg]	6.3 ± 0.1 [ab]	N.D.
D + 28	8.1 ± 0.1 [ghijk]	7.8 ± 0.1 [fgh]	6.1 ± 0.1 [a]	N.D.
Homo-HP	**100 MPa**	**200 MPa**	**300 MPa**	**400 MPa**
D + 1	8.1 ± 0.1 [fghij]	8.0 ± 0.1 [fghij]	7.1 ± 0.2 [d]	N.D.
D + 15	8.1 ± 0.0 [ijk]	7.8 ± 0.3 [f]	6.7 ± 0.1 [c]	N.D.
D + 28	8.1 ± 0.1 [hijk]	7.4 ± 0.1 [e]	6.5 ± 0.1 [bc]	N.D.
Cherry-Flavored Yoghurt	-	-	-	-
Control *	-	-	-	-
D + 1	8.4 ± 0.0 [e]			
D + 15	8.2 ± 0.0 [d]	-	-	-
D + 28	8.1 ± 0.1 [d]			
HP	**100 MPa**	**200 MPa**	**300 MPa**	**400 MPa**
D + 1	8.2 ± 0.0 [d]	7.9 ± 0.0 [c]	6.5 ± 0.1 [a]	N.D.
D + 15	8.1 ± 0.0 [d]	7.7 ± 0.0 [bc]	6.5 ± 0.1 [a]	N.D.
D + 28	8.1 ± 0.1 [d]	7.7 ± 0.1 [bc]	6.5 ± 0.2 [a]	N.D.
Homo-HP	**100 MPa**	**200 MPa**	**300 MPa**	**400 MPa**
D + 1	8.2 ± 0.1 [d]	7.8 ± 0.0 [c]	6.6 ± 0.2 [a]	N.D.
D + 15	8.2 ± 0.0 [d]	7.8 ± 0.0 [c]	6.4 ± 0.2 [a]	N.D.
D + 28	8.1 ± 0.1 [d]	7.6 ± 0.1 [b]	6.4 ± 0.1 [a]	N.D.

* Control samples were not HP-treated but only homogenized at 10 bar after the break of the coagulum. Letter D indicates the day of production. N.D. indicates microbial load below the acceptable limit of 7.0 $\log_{10}$ CFU/g of the total microbial counts for the starter culture at the end of shelf life. Different letters indicate significant difference ($p < 0.05$) between tested samples according to Duncan's mean values post hoc comparison test.

In agreement with the probiotics survival findings in the model system, pressure increase resulted in significantly higher sub-lethality of Bifidobacteria cells compared to Lactobacilli. Bifidobacteria cell population underwent a decrease of about 0.1 to 1.4 $\log_{10}$ CFU/g or 0.1 to 1.1 $\log_{10}$ CFU/g for HP and Homo-HP-treated samples, respectively, when the products treated in the pressure range of 100–300 MPa. The corresponding decrease in Lactobacilli population was 0.1–0.9 $\log_{10}$ CFU/g and 0.0–0.6 $\log_{10}$ CFU/g for HP and Homo-HP-treated samples, respectively. Similar results for probiotics viability were obtained during storage, where a further slight decrease in probiotic populations of ca. 0.2–0.7 $\log_{10}$ CFU/g were observed for all tested samples. Based on these observations it can be hypothesized that after HP treatment there are potentially more injured cells unable to recover in the case of Bifidobacteria than of Lactobacilli. Despite the observed decrease, when products were pressurized in the range of 100–300 MPa, the remaining probiotic population was above the recommended level of 10 [6] CFU/g for both probiotic strains and both treated products (HP and Homo-HP samples) [1]. In the HP-treated products the decrease in probiotic cells was greater than the viability loss observed for Homo-HP-treated products. However, during storage at 5 °C for 28 days, Homo-HP samples presented higher viability loss of probiotic cultures as compared to that of HP samples. Similar to the case of starter culture microorganisms, when products were pressurized at 400 MPa, counts of both probiotic strains decreased below the legislative acceptable levels in all alternatively treated products (depicted as N.D. in Table 4 and, * symbol in Figure 3).

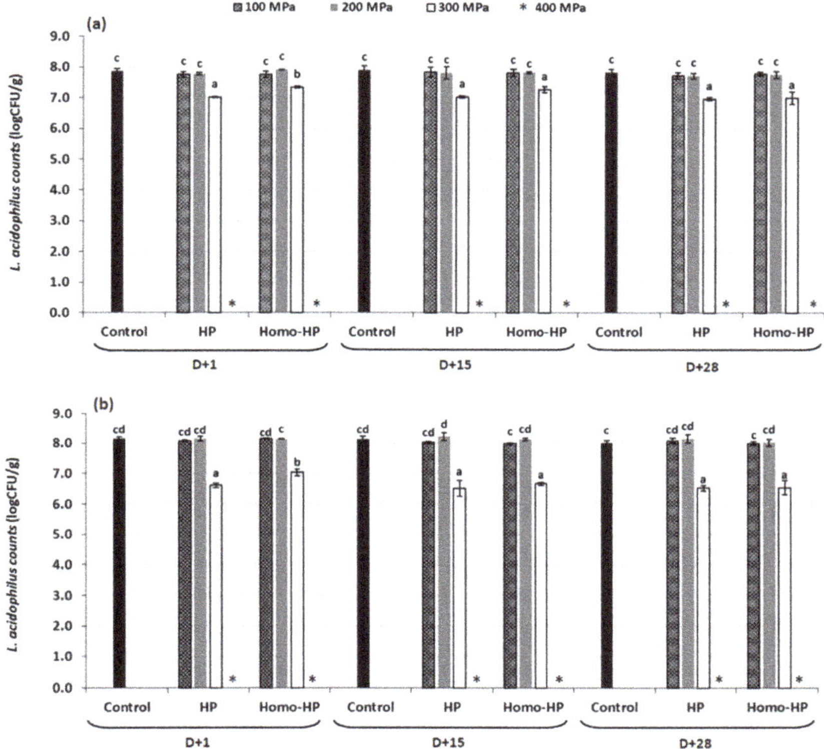

Figure 3. Viability of *Lactobacillus acidophilus* LA5 during 28 days of storage at 5°C in different treated (**a**) plain and, (**b**) cherry-flavored probiotic yoghurt (Letter D indicates the day of production. Different letters among bars indicate significant difference ($p < 0.05$) between tested samples according to Duncan's mean values post hoc comparison test.).

In the case of cherry-flavored yoghurt samples, a similar reduction of about $0.2–0.4 \log_{10}$ CFU/g was observed for total starter culture population when yoghurt samples were subjected to HP processing (both HP and Homo-HP samples), followed by a further decrease of $0.2–0.4 \log_{10}$ CFU/g during 28 days of storage. Probiotic populations in cherry-flavored yoghurt samples subjected to HP processing underwent a reduction of approximately $0.2–1.8 \log_{10}$ CFU/g, followed by a further decrease of 0.2 and $0.1–0.5 \log_{10}$ CFU/g for BB12 and LA5 populations, respectively, during 28 days of storage, indicating a significant lower reduction at the first day of production and during storage to that observed for plain yoghurt samples. Addition of syrup resulted in an increase in the amount of polysaccharides and sugars in these samples, a growth factor for probiotics survival, indicating that the use of substances for the enhancement of taste and flavor in dairy products could also improve the viability of microbial populations [33].

3.3.2. Physicochemical/Rheological Characteristics After HP Treatment and During Storage

The physicochemical (pH, acidity, and WHC) and rheological parameter (consistency coefficient, rheological behavior index and apparent viscosity) data matrices for plain and cherry-flavored yoghurts were averaged, standardized and subjected into PCA (Figure 4).

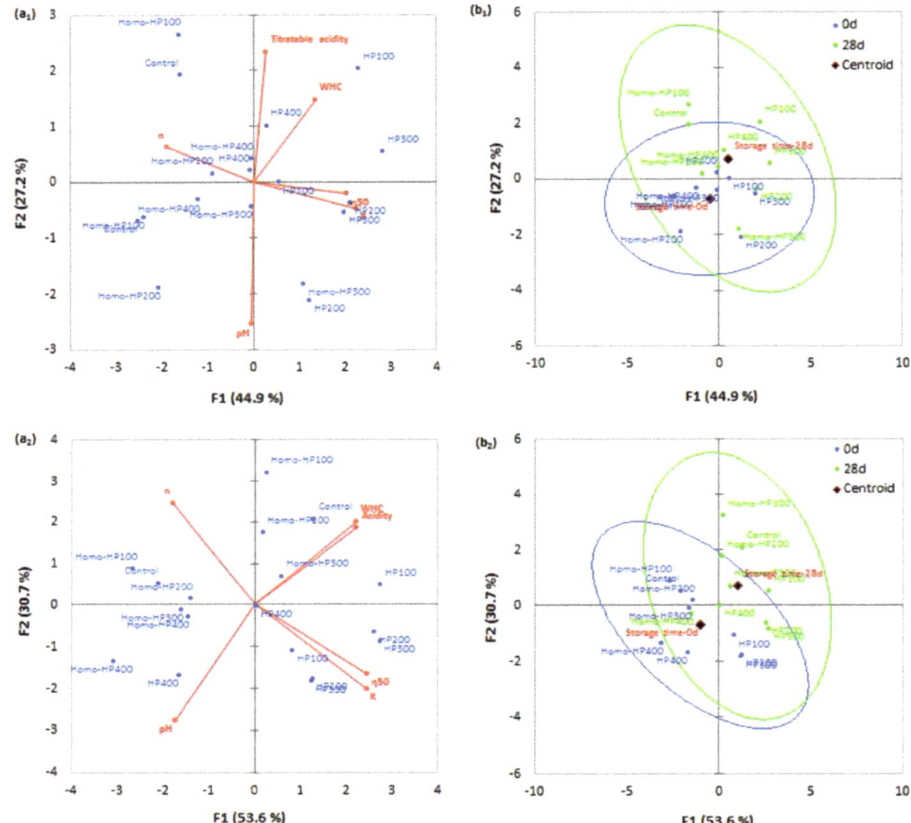

Figure 4. Principal components analysis (a_1,b_1: loadings plot, a_2,b_2: scores plot) for the classification of plain (**a**) and cherry-flavored (**b**) yoghurt samples stored at 5 °C for 28 d, based on their physicochemical characteristics (principal components 1 and 2 accounted together the 72.1% and 84.3% of the total variance explained for plain and cherry-flavored yoghurts, respectively).

As seen in Figure 4, the titratable acidity was reduced proportionally to the intensity of the pressurization process for both plain and cherry-flavored products, which implies that the metabolic activity of the probiotic microorganisms and yoghurt starter is significantly slowed down [34,35]. The post-acidification was found to be pronouncedly higher in the case of cherry-flavored formulations for the entity of the tested processing conditions (HP and Homo-HP) and thus, it is assumed that the presence of readily available nutrient sources in the fruit syrup (e.g., sugar, glucose syrup, natural presence of sugars and fibers in cherry juice) stimulated the growth of the microbiota throughout the chilling storage.

Expectedly the cherry syrup addition suppressed the total color index from 84.5 to 78 for the control systems. A similar trend regarding the impact of the HP processing on total color i.e., from 79.6–85.9 to 75.1–78.9, was observed. However, no significant differences between the HP and Homo-HP samples were identified. Interestingly, the increase in the intensity of pressurization process resulted in a reduction of the total color of cherry-flavored yoghurts, most probably due to the degradation or isomerization of naturally occurring pigmenting compounds such as anthocyanins and carotenoids [36,37].

The WHC is a measure of the ability of acid gel to retain unbound water (loosely hold in the interspaces of the protein gel network) on the application of mechanical stress. It has been demonstrated that the more uniform association between denatured beta-lactoglobulin (as it is more sensitive than alpha-lactalbumin to high-pressure treatment) and the dissociation and re-aggregation of the micelle fragments occurring during the HP processing reduce the proneness of the acid protein gels to syneresis [13]. According to our findings, the WHC in the HP-treated plain yoghurts was significantly higher than that of the Homo-HP-treated ones. On the other hand, a reversed behavior was observed in the case of cherry-flavored yoghurts, where the Homo-HP exerted the highest WHC values. Although there is no conclusive explanation, it is presumed that the breakdown of the acid gel during the homogenization step, allowed the stabilizing agents (e.g., pectins) found in the syrup base, to occupy a higher hydrodynamic volume due to the reduction of steric hindrances, and thus it improved the ability of the overall acid protein/stabilizer protein network to retain more water via hydrogen bond bridging.

The implementation of the HP treatment at the final stage of the production process appeared to improve the rheological properties of the end product, either when it is applied individually or subsequently to conventional homogenization.

The values of consistency coefficient (K) of plain yoghurt beverages were significantly ($p < 0.001$) affected by the applied treatment, the level of the applied pressure and storage time. HP treatment of samples in a pressure range of 100–300 MPa led to an increase in consistency coefficient values compared to the controls (homogenized at 10 bar after the break down of the coagulum). At 400 MPa, consistency coefficient values decreased, although it was still higher than control samples. Flow behavior index values of plain yoghurt beverages were also significantly ($p < 0.001$) affected by the applied treatment and storage time, showing a significant decrease with increasing HP, either applied individually (HP samples) or subsequently to conventional homogenization (Homo-HP samples). This trend was reversed at 400 MPa. The apparent viscosity of plain yoghurt samples increased by 162, 188, 190 and 68% at pressures of 100, 200, 300 and 400 MPa, and 26, 46, 65 and 29% when HP was applied subsequently to conventional homogenization at the same pressures.

The degree of whey protein denaturation is a very important factor that affects the rheological behavior of the coagulum and can be related to the intensity of the pressure applied. It is also responsible for the protein-protein interactions, and the retention of whey proteins in the network gel. HP processing lead to enhanced whey protein hydrophobicity, resulting in an increase in the binding affinity of whey proteins and thus, alterations on their structure and improvement of their functional properties [38,39]. Moreover, it has been reported that when the applied pressure exceeds 200 MPa, a partial disintegration of casein micelles occurs, which results in an increased number of fragments of casein solution and in greater solvation of the protein. The partial fragmentation of casein micelles is accompanied by the solubilization of colloidal calcium phosphate, suggesting that the structure of the gels obtained is dominated by casein-casein interactions instead of the interactions of whey protein-casein, forming small particles that are often shaped into clumps and chains, and therefore into compact and strengthened protein networks [11]. The increase in yoghurt viscosity, with pressure, have been previously related to modifications of beta-lactoglobulin structure, since significant molecular unfolding and further protein aggregation occurs, especially when pressures above 100 MPa are applied and with a threshold at 400 MPa, where usually the coagulum becomes more coarse and an increase in syneresis is observed [16].

The parameters of consistency coefficient and flow behavior index of cherry-flavored yoghurt beverages were affected by the applied treatment similarly to plain yoghurt beverages. However, storage time did not seem to significantly affect these parameters, probably due to the increase in total solids and the presence of small amounts of pectin in the syrup that could have stabilized the structure of the coagulum.

3.3.3. Sensory Profile of the HP Yoghurts

To understand the impact of the processing conditions on the sensory profile of the plain and cherry fruit-flavored yoghurt products, the sensory modalities score data at the beginning and end of the storage time were subjected to PCA (Figure 5). Prior to analyses, the sensory modalities that were not significantly ($p > 0.05$) influenced by the independent processing factors were excluded from the PCA analysis. In this context, one orotactile (grainy), two gustatory (bitter and metallic) and one olfactory (rancid flavor) modality were not affected by processing conditions in both plain and cherry-flavored products. Moreover, white color was not assessed in the case of cherry-flavored yoghurts and accordingly, cherry flavor and cherry fruit color were excluded from the sensory lexicon used in the evaluation of plain yoghurts.

As seen in Figure 5a, the HP processing of the plain yoghurts was primarily associated with the modification of their tactile and oro-tactile aspects. Plain yoghurts pressurized at 100 to 300 MPa exhibited rather a textural and structural affinity in terms of ropiness, coagulum uniformity, thickness, and firmness. Although the increase in the pressure intensity appeared to intensify the aforementioned sense stimuli, the differences between the samples pressurized at 200 and 300 MPa were not significant. On the other hand, further increase in the pressure, i.e., 400 MPa, resulted in colloidally non-uniform, curdy-like products that were prone to gel structural collapse as indicated by the significant evidence of syneresis. As far as concerns the assessed flavor—taste attributes, the increase in the HP process intensity was accompanied by the increase in dairy and partial masking of sourmilk flavor modalities. The observed reverse correlation between dairy/milky and sour/sourmilk or even astringent sense stimuli has been reported in several studied on acidified dairy products [40,41]. Although the amount of organic acids, i.e., lactic, formic or orotic acid, are considered to be the major drivers of the sour-like flavor and taste modalities, the acid protein gel structural conformation can also have a significant role on the partitioning of flavor volatile compounds that contribute to the development of fermented milk olfactory modalities such as acetaldehyde, acetoin, diacetyl, acetone etc. [42,43]. Thus, the HP induced enhancement of the structural integrity of the acid gels can reduce the partitioning coefficients of the aforementioned flavor compounds leading to a more dairy/buttery-like flavor profile.

In the case of the cherry-flavored yoghurts (Figure 5b), the addition of the fruit syrup did not modified remarkably the interplay between the sensory cross-modal perception and the intensity of the HP processing step. As with plain yoghurts, cherry-flavored yoghurts processed at 200–300 MPa exerted the highest intensities of thickness, ropiness, coagulum uniformity and gel firmness and at the same time received the highest scores for cherry fruit flavor and sweet taste modalities. The detrimental effects of the excessive pressure processing on the structural integrity of the coagulum were also observed in the case of the cherry fruit-flavored products.

With regards to the hedonic assessment of the products, the PCA analysis suggested that the DOL is dependent of the product formulation. To further explore the sensory modal drivers of DOL, the standardized and averaged dataset were subjected to PLSR using the leave-one-out validation. As seen in Figure 6a for plain yoghurts, the DOL was mainly driven by tactile/orotactile properties (VIP > 1) and therefore, the structural and colloidal integrity of the formed acid gels has the most important role for the acceptability of the final products. On the other hand, the drivers of overall liking of cherry fruit-flavored yoghurts appeared to be more complex, as not only the texture relating but also the fruit flavor and color attributes classified as impacting attributes (Figure 6b). Finally, the storage time had only minor ($p > 0.05$) effects on the sensorial quality and overall liking of both type of yoghurts, most probably due to the very mild post-acidification and colloidal change phenomena occurring throughout the tested storage period.

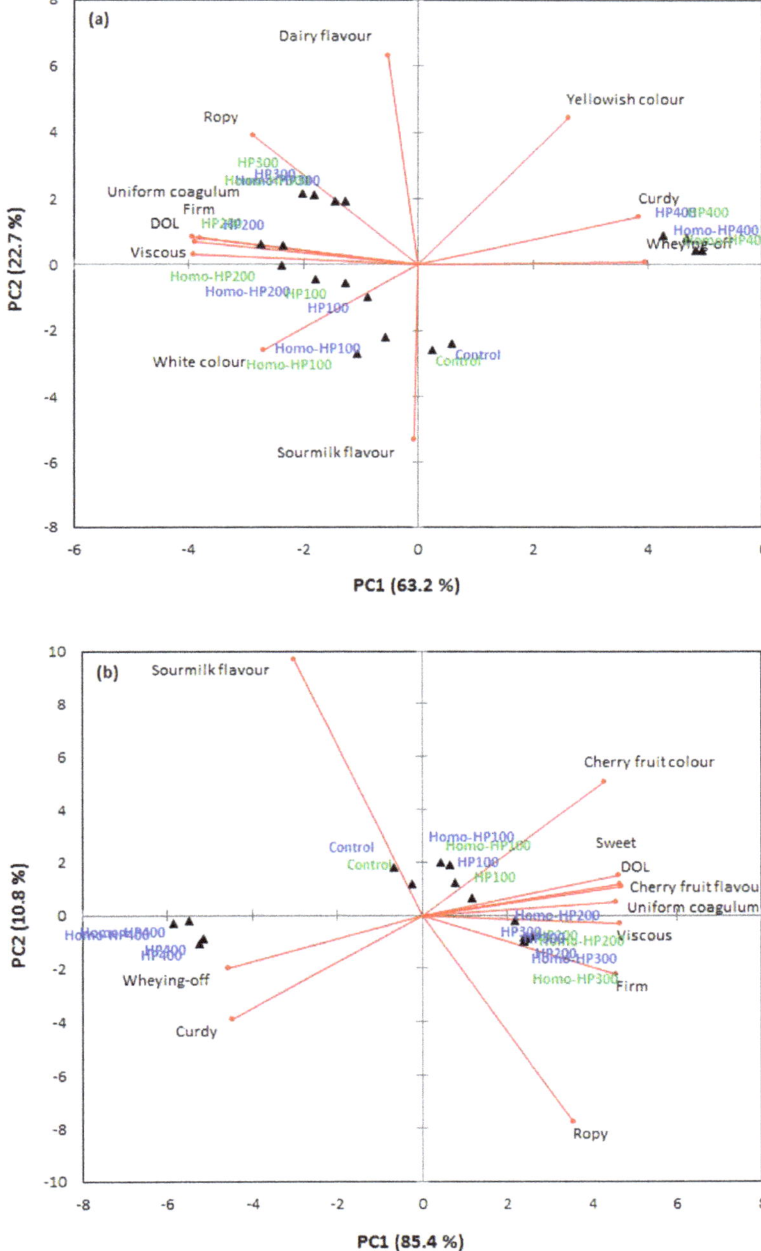

Figure 5. Principal components analysis (**a**) plain yoghurt, (**b**) cherry-flavored yoghurt for the classification of the yoghurt samples at 0d (blue) and 28d (green) based on their sensory characteristics (principal components 1 and 2 accounted together the 96.2% of the total variance explained).

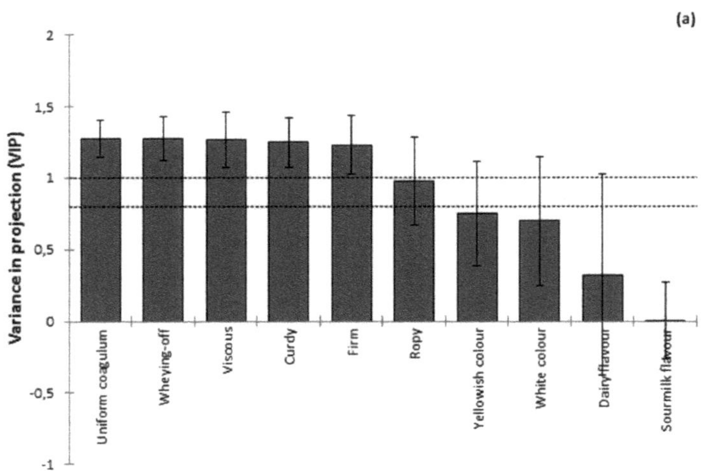

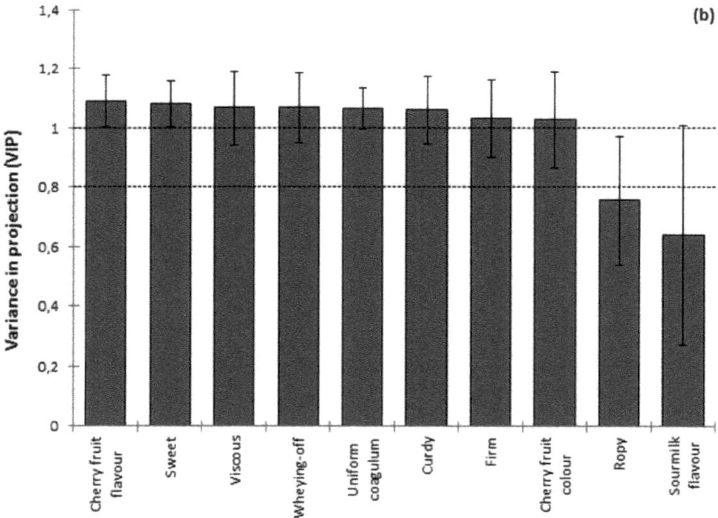

Figure 6. Variable in Project values (VIP) in descending order as calculated according to the partial least squares regression (PLSR) analysis. (**a**) VIP calculated for plain yoghurts according to PLSR analysis; (**b**) VIP calculated for cherry fruit-flavored yoghurts according to PLSR analysis.

4. Conclusions

The recommended HP conditions in the literature (200 MPa at 20–25 °C for process time for up to 10–15 min) for texture improvement in dairy products was found not detrimental to the viability of the examined probiotic bacteria. In the case of model systems, the maximum calculated inactivation due to HP process was about 0.4 $\log_{10}$ CFU/mL. To describe the effect of pressure and temperature process conditions on the viability of the tested probiotic bacteria, a single multi-parameter equation was proposed, and the parameters of this model were estimated for the most baro-sensitive probiotic bacterium (i.e., *Bifibobacterium bifidum*). When HP processing was applied in probiotic yoghurt at the final stage of production, especially when treated at 200–300 MPa, the quality and sensorial properties of the final product were improved. The viscosity of the final product increased and the whey separation decreased, while the viability loss of the probiotic microorganisms ranged between 0.5–1.2 $\log_{10}$ CFU/g, and no significant viability loss was observed during refrigerated storage for 28 days. Overall, HP process can be successfully applied in such dairy products at the final step of production in order to improve their quality attributes and extend their shelf life without the need for stabilizer addition, while acceptably affecting their functionality. Moreover, the addition of flavor enhancement substances seems to improve the rheological properties of the coagulum but also increase the viability of the probiotic bacteria. Further research could focus on the effect of the addition of prebiotics in similar treated dairy products both on probiotics viability and their quality indices.

Author Contributions: Conceptualization, M.T. and P.T.; methodology, M.T. and C.S.; validation, M.T. and C.S.; formal analysis, M.T., C.S. and M.O.-R.; investigation, M.T., M.O.-R. and C.S.; writing—original draft preparation, M.T. and C.S.; visualization, M.T. and C.S.; supervision, M.T. and P.T.; project administration, P.T. All authors have read and agreed to the published version of the manuscript.

Funding: This research was co-funded by STATE SCHOLARSHIPS FOUNDATION of Greece, in the context of Dr Maria's Tsevdou PhD Scholarship Grant (Food Engineering, 0840620729.006.031).

Acknowledgments: The authors would like to kindly thank Christian–Hansen Hellas for the gift of *Bifidobacterium lactis* BB12 and *Lactobacillus acidophilus* LA5 probiotic cultures.

References

1. Food and Agriculture Organization of the United Nations/World Health Organization (FAO/WHO). *Report of a Joint FAO/WHO Working Group on Drafting Guidelines for the Evaluation of Probiotics in Food*; Food and Agriculture Organization of the United Nations/World Health Organization: London, ON, Canada, 2002.

2. Hill, C.; Guarner, F.; Reid, G.; Gibson, G.R.; Merenstein, D.J.; Pot, B.; Morelli, L.; Canani, R.B.; Flint, H.J.; Salminen, S.; et al. The International Scientific Association for Probiotics and Prebiotics consensus statement on the scope and appropriate use of the term probiotic. *Nat. Rev. Gastroenterol. Hepatol.* **2014**, *11*, 506–514. [CrossRef]

3. Şanlier, N.; Gökcen, B.B.; Sezgin, A.C. Health benefits of fermented foods. *Crit. Rev. Food Sci. Nutr.* **2017**, *59*, 506–527. [CrossRef]

4. Collado, M.C.; Isolauri, E.; Salminen, S.; Sanz, Y. The impact of probiotic on gut health. *Curr. Drug. Metab.* **2008**, *10*, 68–78. [CrossRef] [PubMed]

5. Tripathy, M.K.; Giri, S.K. Probiotic functional foods: Survival of probiotics during processing and storage. *J. Funct. Foods* **2014**, *9*, 225–241. [CrossRef]

6. Lucey, J.A. Formation and physical properties of milk protein gels. *J. Dairy Sci.* **2002**, *85*, 281–294. [CrossRef]

7. Leroy, F.; De Vuyst, L. Lactic acid bacteria as functional starter cultures for the food fermentation industry. *Trends Food Sci. Technol.* **2004**, *15*, 67–78. [CrossRef]

8. Trujillo, A.J.; Capellas, M.; Saldo, J.; Gervilla, R.; Guamis, B. Application of high-hydrostatic pressure on milk and dairy products: A review. *Innov. Food Sci. Emerg.* **2002**, *3*, 295–307. [CrossRef]

9. Anema, S.G.; Lauber, S.; Lee, S.K.; Henle, T.; Klostermeyer, H. Rheological properties of acid gels prepared from pressure- and transglutaminase-treated skim milk. *Food Hydrocoll.* **2006**, *19*, 879–887. [CrossRef]

10. Penna, A.L.B.; Subbarao-Gurram, G.V.; Barbosa-Cánovas, G.V. High hydrostatic pressure processing on microstructure of probiotic low-fat yogurt. *Food Res. Int.* **2007**, *40*, 510–519. [CrossRef]

11. Masson, L.M.P.; Rosenthal, A.; Calado, V.M.A.; Dliza, R.; Tashima, L. Effect of ultra-high pressure homogenization on viscosity and shear stress of fermented dairy beverage. *LWT-Food Sci. Technol.* **2011**, *44*, 495–501. [CrossRef]

12. Tsevdou, M.S.; Eleftheriou, E.G.; Taoukis, P.S. Transglutaminase treatment of thermally and high pressure processed milk: Effects on the properties and storage stability of set yoghurt. *Innov. Food Sci. Emerg.* **2013**, *17*, 144–152. [CrossRef]

13. Loveday, S.M.; Sarkar, A.; Singh, H. Innovative yoghurts: Novel processing technologies for improving acid milk gel texture. *Trends Food Sci. Technol.* **2013**, *33*, 5–20. [CrossRef]
14. Tanaka, T.; Hatanaka, K. Application of hydrostatic pressure to yoghurt to prevent its after-acidification. *J. Jpn. Soc. Food Sci.* **1992**, *39*, 173–177. [CrossRef]
15. Krompkamp, J.; Moreira, R.M.; Langeveld, L.P.M.; Van Mil, P.J.J.M. Microorganisms in milk and yoghurt: Selective inactivation by high hydrostatic pressure. In Proceedings of the IDF Symposium Heat Treatments and Alternative Methods, Vienna, Austria, 6–8 September 1995.
16. de Ancos, B.; Cano, M.P.; Gómez, R. Characteristics of stirred low-fat yoghurt as affected by high pressure. *Int. Dairy J.* **2000**, *10*, 105–111. [CrossRef]
17. Tarmaraj, N.; Shah, N.P. Selective Enumeration of Lactobacillus delbrueckii ssp. bulgaricus, Streptococcus thermophilus, Lactobacillus acidophilus, Bifidobacteria, Lactobacillus casei, Lactobacillus rhamnosus and Propionibacteria. *J. Dairy Sci.* **2003**, *86*, 2288–2296. [CrossRef]
18. International Dairy Federation (IDF). *Yogurt: Determination of Titratable Acidity. International IDF Standard, ISO 1991*; International Dairy Federation: Vienna, Austria, 1991.
19. Harte, F.; Luedecke, L.; Swanson, B.; Barbosa-Cánovas, G.V. Low-fat set yogurt made from milk subjected to combinations of high hydrostatic pressure and thermal processing. *J. Dairy Sci.* **2003**, *86*, 1074–1082. [CrossRef]
20. Akhtar, M.; Murray, B.S.; Dickinson, E. Perception of creaminess of model oil-in-water dairy emulsions: Influence of the shear-thinning nature of a viscosity-controlling hydrocolloid. *Food Hydrocoll.* **2006**, *20*, 839–847. [CrossRef]
21. ISO 13300-1. *Sensory Analysis—General Guidance for the Staff of a Sensory Evaluation Laboratory—Part 1: Staff Responsibilities*; ISO: Geneva, Switzerland, 2006.
22. ISO 13300-2. *Sensory Analysis—General Guidance for the Staff of a Sensory EVALUATION Laboratory—Part 2: Recruitment and Training of Panel Leaders*; ISO: Geneva, Switzerland, 2006.
23. ISO 8589. *Sensory Analysis—General Guidance for the Design of Test Rooms*; ISO: Geneva, Switzerland, 2007.
24. Katsaros, G.I.; Tsevdou, M.; Panagiotou, T.; Taoukis, P.S. Kinetic study of high pressure microbial and enzyme inactivation and selection of pasteurization conditions for Valencia Orange Juice. *Int. J. Food Sci. Technol.* **2010**, *45*, 1119–1129. [CrossRef]
25. Kanta, A.; Soukoulis, C.; Tzia, C. Eliciting the Sensory Modalities of Fat Reformulated Yoghurt Ice Cream Using Oligosaccharides. *Food Bioprocess Technol.* **2018**, *11*, 885–900. [CrossRef]
26. Dave, R.I.; Shah, N.P. Effect of Cysteine on the Viability of Yoghurt and Probiotic Bacteria in Yoghurts Made with Commercial Starter Cultures. *Int. Dairy J.* **1997**, *7*, 537–545. [CrossRef]
27. Gomez, A.; Malcata, F. *Bifidobacterium* spp. and *Lactobacillus acidophilus*: Biological, biochemical, technological and therapeutical properties relevant for use as a probiotic. *Trends Food Sci. Technol.* **1999**, *10*, 139–157. [CrossRef]
28. Reyns, K.M.F.A.; Sootjens, C.C.F.; Cornelis, K.; Weemars, C.A.; Hendrickx, M.E.; Michiels, C.W. Kinetics analysis and modeling of combined high–pressure–temperature inactivation of the yeast Zygosaccharomyces bailii. *Int. J. Food Microbiol.* **2000**, *56*, 199–210. [CrossRef]
29. Godwart, G.; Kailasapathy, K. Viability and survival of free, encapsulated probiotic bacteria in yoghurt. *Milchwissenschaft* **2003**, *58*, 396–399.
30. Yilmaztekin, M.; Özer, B.H.; Atasoy, F. Survival of *Lactobacillus acidophilus* LA-5 and *Bifidobacterium bifidum* BB-02 in white-brined cheese. *Int. J. Food Sci. Nutr.* **2004**, *55*, 53–60. [CrossRef]
31. Özer, B.; Kirmaci, H.A.; Şenel, E.; Atamer, M.; Hayaloğlu, A. Improving the viability of *Bifidobacterium bifidum* BB-12 and *Lactobacillus acidophilus* LA-5 in white-brined cheese by microencapsulation. *Int. Dairy J.* **2009**, *19*, 22–29. [CrossRef]
32. Food and Agriculture Organization of the United Nations/World Health Organization (FAO/WHO). *Standard for Fermented Milks, Revised Standard. Codex Alimentarius. STAN 243-2003*; Food and Agriculture Organization of the United Nations/World Health Organization: Rome, Italy, 2011.
33. Senaka Ranadheera, C.; Evans, C.A.; Adams, M.C.; Baines, S.K. Probiotic viability and physico-chemical and sensory properties of plain and stirred fruit yogurts made from goat's milk. *Food Chem.* **2012**, *135*, 1411–1418. [CrossRef]
34. Jankowska, A.; Grześkiewicz, A.; Wiśniewska, K.; Reps, A. Application of probiotic bacteria in production of yoghurt preserved under high pressure. *High Press. Res* **2005**, *25*, 57–62. [CrossRef]

35. Jankowska, A.; Grześkiewicz, A.; Wiśniewska, K.; Reps, A. Examining the possibilities of applying high pressure to preserve yoghurt supplemented with probiotic bacteria. *High Press. Res.* **2012**, *32*, 339–346. [CrossRef]

36. Oey, I.; Lille, M.; Van Loey, A.; Hendrickx, M. Effect of high pressure processing on colour, texture and flavour of fruit and vegetable-based food products: A review. *Trends Food Sci. Technol.* **2008**, *19*, 320–328. [CrossRef]

37. Barba, F.J.; Esteve, M.J.; Frigola, A. Physicochemical and nutritional characteristics of blueberry juice after high pressure processing. *Food Res. Int.* **2008**, *50*, 545–549. [CrossRef]

38. Liu, X.; Powers, J.R.; Swanson, B.G.; Hill, H.H.; Clark, S. Modification of whey protein concentrate hydrophobicity by high hydrostatic pressure. *Innov. Food Sci. Emerg.* **2005**, *6*, 310–317. [CrossRef]

39. Lee, W.; Clark, S.; Swanson, B.G. Functional properties of high hydrostatic pressure-treated whey protein. *J. Food Process. Preserv.* **2006**, *30*, 488–501. [CrossRef]

40. Soukoulis, C.; Lyroni, E.; Tzia, C. Sensory profiling and hedonic judgement of probiotic ice cream as a function of hydrocolloids, yogurt and milk fat content. *LWT-Food Sci. Technol.* **2010**, *43*, 1351–1358. [CrossRef]

41. Lesme, H.; Rannou, C.; Famelart, M.-H.; Bouhallab, S.; Prost, C. Yogurts enriched with milk proteins: Texture properties, aroma release and sensory perception. *Trends Food Sci. Technol.* **2020**. [CrossRef]

42. Tsevdou, M.; Soukoulis, C.; Cappellin, L.; Gasperi, F.; Taoukis, P.S.; Biasioli, F. Monitoring and Modeling of Endogenous Flavour Compounds Evolution during Fermentation of Thermally, High Hydrostatic Pressure or Transglutaminase Treated Milk using PTR-TOF-MS. *Food Chem.* **2013**, *138*, 2159–2167. [CrossRef]

43. Soukoulis, C.; Cappellin, L.; Aprea, E.; Costa, F.; Viola, R.; Märk, T.D.; Gasperi, F.; Biasioli, F. PTR-ToF-MS, A Novel, Rapid, High Sensitivity and Non-Invasive Tool to Monitor Volatile Compound Release During Fruit Post-Harvest Storage: The Case Study of Apple Ripening. *Food Bioprocess. Technol.* **2013**, *6*, 2831–2843. [CrossRef]

Article

Microfiltration of Ovine and Bovine Milk: Effect on Microbial Counts and Biochemical Characteristics

George Panopoulos, Golfo Moatsou, Chrysanthi Psychogyiopoulou and Ekaterini Moschopoulou *

Laboratory of Dairy Research, Department of Food Science and Human Nutrition, Agricultural University of Athens, Iera Odos 75, 118 55 Athens, Greece; g.panopoulos@yahoo.gr (G.P.); mg@aua.gr (G.M.); ps.chrysanthi@gmail.com (C.P.)
* Correspondence: catmos@aua.gr

Received: 9 February 2020; Accepted: 2 March 2020; Published: 4 March 2020

Abstract: The aim of this research work was to assess the effect of the microfiltration (ceramic membranes 1.4 μm, 50 °C) of partially defatted ovine milk (fat 0.4%) and bovine milk (fat 0.3%) characteristics. Feed milks, permeates and retentates were analyzed for microbial counts, gross composition, protein fractions, the indigenous enzymes cathepsin D and alkaline phosphatase and the behavior during renneting. It was showed that the microbial quality of both ovine and bovine permeate was improved by reduction of the total mesophilic microflora about 4 Log and 2 Log, respectively. The protein contents and the total solids contents of both permeates were significantly ($p < 0.05$) reduced. A further analysis of protein fractions by Reversed Phase -High Performance Liquid Chromatography (RP-HPLC) revealed lower αs_1- and β-casein and higher κ-casein contents in permeates. The activity of alkaline phosphatase followed the allocation of the fat content, while activity of cathepsin D in permeates was not influenced, although somatic cells counts were removed. Regarding cheesemaking properties, the firmness of ovine curd made from the feed milk did not differ significantly from that made from the permeate. The obtained results suggested that microfiltration could be used for pre-treating of ovine milk prior to cheesemaking.

Keywords: microfiltration; ovine milk; bovine milk; casein fractions; alkaline phosphatase; cathepsin D; milk renneting properties

---

1. Introduction

The membrane process is widely used in the dairy industry for separation or fractionation purposes, depending on the membrane pore size and the applied pressure [1,2]. Microfiltration (MF) involves membranes with a pore size of 0.1–10 μm and operates at pressures of 0.1–8 bars [3,4]. MF is mainly applied to reduce bacteria, spores and somatic cells [1,4–7] in fluid dairy products extending, in this way, their shelf life [8,9]. Therefore, MF combined with pasteurization is an excellent process to produce Extended Shelf Life (ESL) milk that is considered 'purer' and more 'natural' than standard heat-treated milk. Moreover, MF can be used to fractionate globular milk fat [10], to fractionate and concentrate casein or to purify β-Lactoglobulin (β-Lg) [5,11], while in combination with ultrafiltration, it is used in whey processing. In addition, MF has been studied in regards to pre-treatment of milk for cheese production [12]. All above applications, which are focused on bovine milk, require membranes of different pore sizes. For bacteria reduction, the most used membranes are those of pore size 1.4 μm, since it has been shown that MF of bovine milk through a membrane of 1.4 μm pore size effectively reduces the microbial populations without significantly affecting the major milk components [13,14].

Raw ovine milk is richer in total solids and contains higher microbial counts than bovine milk. According to European Community regulation, the microbial criteria for ovine milk define that the two-months rolling geometric average of microbes (with at least two samples per month) grown on

Plate Count Agar at 30 °C should be a maximum 1,500,000 per mL. [15]. To the best of our knowledge, studies for MF of ovine milk are rare and concern only the impact on microflora [16,17]. Moreover, there is lack of information about the influence of MF on indigenous enzymes or on the renneting behavior of either ovine or bovine milk. Therefore, the aim of this study is to evaluate the effect of MF, using a ceramic membrane of 1.4 μm pore size, on the composition, microflora, cathepsin D and alkaline phosphatase as well as on cheesemaking properties of ovine and bovine milk.

2. Materials and Methods

2.1. Microfiltration

Raw ovine and bovine milk obtained from the farm of Agricultural University of Athens were defatted to fat content 0.30% and 0.40%, respectively, using a laboratory milk fat separator, and were subsequently heated at 50 °C. Crossflow MF processing (Figure 1) took place on a pilot microfiltration module (PALL Italia s.r.l, MILANO, Italy) under a constant transmembrane pressure (TMP) of 1.5 bar at 50 °C using ceramic membrane P19–40 (Membralox®) of 1.4 μm pore size, with length 1020 mm and area 0.24 m². Filtration duration was 15 min with constant flux at 105 ± 32 L·m^{-2}·h^{-1} for ovine milk and 314 ± 32 L·m^{-2}·h^{-1} for bovine milk. Retention flow rates were 363.23 ± 86.32 L·m^{-2}·h^{-1} and 795.83 ± 18.37 L·m^{-2}·h^{-1} for bovine and ovine milk retentates, respectively. After processing, the microfiltration unit was cleaned by rinsing in series with water, NaOH 1% solution, water, HNO$_3$ 1% solution and water to reach pH 7.0 [17]. Five and three experimental trials were carried out for ovine milk and bovine milk, respectively. Samples codes were as follows: feed ovine milk (O), permeate ovine milk (OP), retentate ovine milk (OR), feed bovine milk (B), permeate bovine milk (BP) and retentate bovine milk (BR).

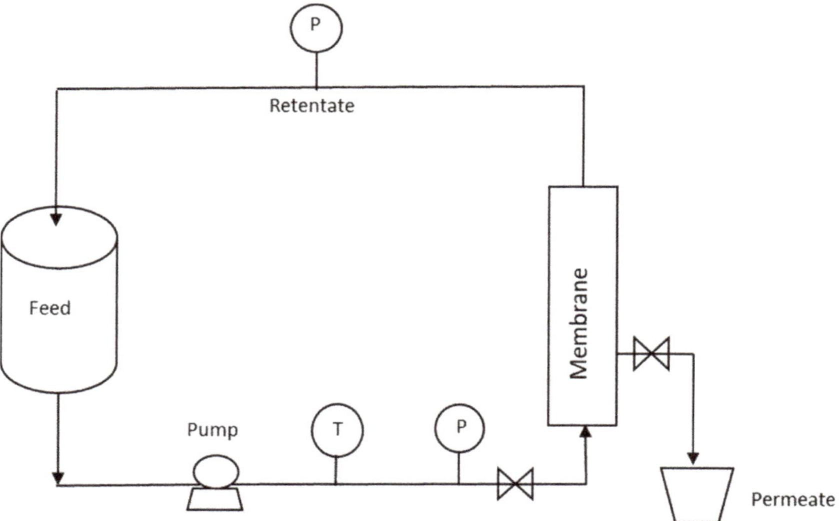

Figure 1. Simplified diagram showing the crossflow microfiltration process.

2.2. Physicochemical Analyses

The pH was measured on a pH meter and acidity was determined by the titration method using 0.11 N NaOH solution. Fat, protein, lactose and total solids contents were determined by means of infrared spectroscopy (Milkoscan FT6000, Foss, Hillerod, Denmark). Ash content was determined by the AOAC method [18]. Total nitrogen (TN) as well as water soluble nitrogen (WSN) were determined by the Kjeldahl method. Phosphorus content was determined by molecular absorption spectrometry [19]. Moreover, calcium content of ovine milk was measured by Atomic Absorption Spectrometry [20] on a Shimadzu AA-6800 Atomic Absorption Spectrophotometer (Shimadzu, Kyoto, Japan) equipped with the autosampler Shimadzu ASC-6100 and the software WizAArd v. 2.30.

Somatic cell counts (SCC) were determined on Fossomatic (Foss, Hillerod, Denmark). All analyses were performed in duplicate.

2.3. Protein Composition

Casein fractions and whey proteins were determined by the Reversed Phase -High Performance Liquid Chromatography (RP-HPLC) method [21] on a HPLC system consisting of a pump capable of mixing four solvents (Waters 600E, Waters, Milford, MA, USA), a photodiode array detector (Waters 996), a helium degasser, a Rheodyne 7125 injector (Rheodyne Inc., Cotati, CA, USA) and Millennium software (v.3.05.01, Waters, Milford, MA, USA).

2.4. Microbial Analyses

The populations of different microbial groups were estimated by the pour plate method using selective growth media and incubation conditions as follows: total mesophilic microflora on Plate Count Agar at 30 °C for 72 h; coliforms on Violet Red Bile Agar at 37 °C for 24 h; thermophilic lactococci on M17 Agar at 42 °C for 48 h; thermophilic lactobacilli on MRS Agar at 42 °C for 72 h; yeasts and molds on Yeast Glucose Chloramphenicol Agar at 25 °C for 72 h; aerobic spore forming microorganisms on Soluble Nutrition Agar at 42 °C for 48 h; anaerobic spore forming microorganisms on Reinforced Clostridia Medium at 42 °C for 14 days anaerobically.

2.5. Indigenous Enzymes Activity

2.5.1. Cathepsin D Activity

Cathepsin D activity was determined by the HPLC method proposed by O'Drissol et al. [22] and modified by Hurley et al. [23] using the synthetic peptide Pro-Thr—Glu-Phe-[p-nitro-Phe]-Arg-Leu (Bachem, Switzerland) as substrate. The Waters HPLC system described previously was used. Sample preparation, HPLC conditions and results interpretation were according to Moatsou et al. [24].

2.5.2. Alkaline Phosphatase Activity

Alkaline Phosphatase (ALP) activity was determined according to the spectrophotometric method of the International Dairy Federation Standard [25]. The milk sample was diluted with a buffer, pH 10.6, containing disodium phenyl phosphate (substrate) and was incubated at 37 °C for 1 h. After incubation, any active ALP had liberated phenol from the substrate. The phenol was detected by adding 2,6-dibromoquinonechlorimide to detect blue color (dibromoindophenol), which was measured on a spectrophotometer (model Lambda 20, Perkin Elmer, Waltham, MA, USA) at 610 nm. The ALP activity was expressed in μg of phenol per mL of milk.

2.6. Rennet Clotting Behavior

The behavior of MF permeates and retentates during clotting with rennet (Nature extra 1125; Hansen Denmark) and was determined on a Formagraph (Lattoninamograph; Foss, Padova, Italia). Milk clotting time and curd firmness after 30 min were recorded.

2.7. Statistical Analysis

The obtained data were subjected to statistical analysis using the software Statgraphics (Statistical Graphics Corp. Rockville, Maryland, MD, USA). Comparisons of means were made using the Least Significant Difference test (LSD, $p < 0.05$).

3. Results and Discussion

3.1. Physicochemical Composition

The gross composition, pH value and acidity of both ovine and bovine milk are shown in Table 1. First, as feed milks were not skim, i.e., 0.05% fat, a significant ($p < 0.05$) reduction of fat contents of permeates was observed. This was expected since MF, even through a membrane of 1.4 μm pore diameter, removes fat [4]. Second, protein content was significantly ($p < 0.05$) lower in both ovine and bovine permeate than in the respective feed milks O and B. This was probably due to the significant lower casein nitrogen (Table 2). In contrast, the water-soluble nitrogen components did not exhibit a significant difference between permeates and retentates of both milks. The casein nitrogen to total nitrogen (CN/TN) ratio showed 1.3% and 5.1% retention of casein for bovine and ovine permeate, respectively, despite the similar average diameter of casein micelles, i.e., 193 nm in ovine and 180 nm in bovine milk [26]. Because MF operates at relatively low pressures, it is susceptible to fouling and to formation of a secondary layer on the membrane by gelatinous material when higher pressure and fluxes are applied [3]. In addition, the length of the tubular membrane that was used was long enough (1.02 m) for the formation of a secondary layer, since it has been shown that tubular membranes 1.2 m long usually operate under a deposit layer [27]. The flux in this experiment was constant during the 15 min of MF process. Therefore, the retention of casein micelles could be due to the formation of a secondary membrane on membrane surface. Other researchers have reported non-significant casein retention during MF (1.4 μm) of bovine milk [13,28].

Table 1. Effect of microfiltration on pH, acidity, somatic cells counts (SCC) and chemical composition (%) of partially defatted bovine (B) or ovine (O) milk, bovine permeate (BP), bovine retentate (BR) (mean ± SD, n = 3), ovine permeate (OP) and ovine retentate (OR) (mean ± SD, n = 5).

	Bovine			Ovine		
	B	**BP**	**BR**	**O**	**OP**	**OR**
pH	6.68 ± 0.02	6.67 ± 0.03	6.68 ± 0.02	6.60 ± 0.07	6.53 ± 0.07	6.49 ± 0.08
Acidity	0.15 ± 0.02 [a,*]	0.13 ± 0.00 [b]	0.14 ± 0.01 [a,b]	0.22 ± 0.02 [a]	0.19 ± 0.02 [b]	0.22 ± 0.02 [a]
Fat	0.29 ± 0.10 [a,b]	0.05 ± 0.02 [a]	0.46 ± 0.19 [b]	0.41 ± 0.09 [a]	0.16 ± 0.01 [b]	0.43 ± 0.11 [a]
Protein	3.43 ± 0.15 [a]	3.03 ± 0.21 [b]	3.63 ± 0.15 [a]	5.71 ± 0.28 [a]	4.68 ± 0.17 [b]	5.78 ± 0.28 [a]
Lactose	4.91 ± 0.12	4.96 ± 0.23	4.96 ± 0.22	4.83 ± 0.05	4.76 ± 0.04	4.80 ± 0.05
Total Solids	8.88 ± 0.29 [a,b]	8.3 ± 0.21 [a]	9.24 ± 0.44 [b]	10.87 ± 0.46 [a]	9.32 ± 0.21 [b]	10.95 ± 0.46 [a]
Ash	0.78 ± 0.01 [a]	0.75 ± 0.02 [b]	0.81 ± 0.02 [a]	0.94 ± 0.04 [a]	0.82 ± 0.03 [b]	0.96 ± 0.04 [a]
Phosphorus (mg/100 g)	105.35 ± 5.73 [a]	97.54 ± 1.43 [b]	107.19 ± 1.65 [a]	149.62 ± 5.46 [a]	127.85 ± 5.50 [b]	154.70 ± 8.36
SCC	417500.00 ± 43100.00 [a]	0	653000.00 ± 73500.00 [b]	660400.00 ± 199600.00 [a]	0	570600.00 ± 159000.00 [b]

* For the same milk, means with different superscript in the same row differ significantly ($p < 0.05$).

The total solids content of permeates were similarly affected by the casein retention (Table 1). Casein retention affects also the phosphorus and calcium contents, since part of these inorganic elements are associated with casein micelles. Because of this, the mean phosphorus content of bovine permeate milk was significantly lower than of bovine feed milk and its retentate. The same trend was observed for the mean phosphorus content of ovine permeate milk. Moreover, the mean calcium content of ovine permeate milk was 148.56 ± 7.77 mg/100g, and it was significantly ($p < 0.05$) lower than in feed milk (198.01 ± 15.86 mg/100g) and retentate (191.06 ± 6.94 mg/100g). Ash contents were like phosphorus and calcium contents, i.e., significantly lower in ovine and bovine permeates than in the respective feed milk (Table 1).

Table 2. Effect of microfiltration on the nitrogen components (%) of partially defatted bovine (B) or ovine (O) milk, bovine permeate (BP), bovine retentate (BR) (mean ± SD, n = 3), ovine permeate (OP) and ovine retentate (OR) (mean ± SD, n = 5).

	Bovine			Ovine		
	B	**BP**	**BR**	**O**	**OP**	**OR**
Total nitrogen (TN)	0.55 ± 0.03 [a]	0.48 ± 0.02 [b]	0.55 ± 0.02 [a]	0.88 ± 0.04 [a]	0.71 ± 0.02 [b]	0.90 ± 0.03 [a]
Casein nitrogen (CN)	0.43 ± 0.04 [a]	0.35 ± 0.02 [b]	0.43 ± 0.03 [a]	0.67 ± 0.03 [a]	0.51 ± 0.04 [b]	0.68 ± 0.03 [a]
Water soluble nitrogen	0.12 ± 0.01	0.13 ± 0.01	0.13 ± 0.02	0.21 ± 0.04	0.20 ± 0.03	0.22 ± 0.04
CN/TN	0.75 ± 0.01	0.74 ± 0.04	0.75 ± 0.07	0.78 ± 0.01 [a]	0.74 ± 0.02 [a]	0.78 ± 0.0 [b]

For the same milk, means with different superscript in the same row differ significantly ($p < 0.05$).

Finally, pH values were not affected, but acidities of both OP and BP were significantly ($p < 0.05$) lower than in feed milks. Milk acidity is related to protein content and, thus, the lower acidity values of permeates could be associated with the lower protein content that both permeates presented.

3.2. Protein Composition

Further analysis of proteins by RP-HPLC revealed the results shown in Table 3. It is obvious that the application of crossflow MF, even with a membrane of 1.4 μm pore size, resulted in the fractionation of casein micelles, since protein permeation depends on the interaction between membrane pore and protein size. Both ovine and bovine milk permeates had significantly ($p < 0.05$) higher κ-casein (κ-CN) contents and lower β-casein (β-CN) contents than their counterpart's retentates. Similar, but not statistically significant, results have been reported by Tziloula et al. [28], who showed that micelles of bovine milk with a diameter greater than 550 nm were retained during MF (1.4μm). In ovine permeate, except for β-CN, the α_{s1}-CN and α_{s2}-CN contents were significantly ($p < 0.05$) lower. The loss of casein fractions and especially those of κ-CN in retentate may be due to the damage of the casein micelles surface, which is caused by the shear forces in the membrane circuit pump [11]. In the case of ovine milk, the shear forces were lower because of the lower flux compared to bovine milk and, thus, more κ-CN remained in permeate. On the other hand, α-lactalbumin (α-La) and β-Lg were concentrated in the permeates because of their smaller volume. The particle diameter of bovine α-La and β-Lg was less than 10 nm [29].

Table 3. Effect of microfiltration on casein fractions (% of total protein) and the main whey proteins (% of total protein) of partially defatted bovine (B) or ovine (O) milk, bovine permeate (BP), bovine retentate (BR) (mean ± SD, n = 3), ovine permeate (OP) and ovine retentate (OR) (mean ± SD, n = 5).

	Bovine			Ovine		
	B	**BP**	**BR**	**O**	**OP**	**OR**
κ-CN	11.02 ± 0.30 [a]	11.47 ± 0.48 [b]	11.56 ± 0.24 [b]	9.71 ± 0.43 [a]	10.14 ± 0.51 [b]	9.77 ± 0.17 [a]
αs1-CN	28.62 ± 0.82	28.19 ± 0.60	28.90 ± 1.26	29.02 ± 0.81 [a]	27.42 ± 0.53 [b]	29.41 ± 0.82 [a]
αs2-CN	8.96 ± 0.77	9.03 ± 0.63	8.93 ± 0.83	12.51 ± 0.95 [a]	11.66 ± 0.88 [b]	12.58 ± 0.70 [a]
β-CN	34.09 ± 0.77 [a]	32.59 ± 0.58 [b]	33.11 ± 1.00 [b]	32.01 ± 1.66 [a]	30.67 ± 1.03 [b]	31.83 ± 0.85 [a]
α-la	2.69 ± 0.28 [a,b]	2.91 ± 0.33 [a]	2.52 ± 0.23 [b]	3.72 ± 0.24 [a]	4.41 ± 0.28 [b]	3.80 ± 0.46 [a]
β-lg	8.07 ± 0.71	8.49 ± 0.91	7.75 ± 0.65	8.11 ± 0.57 [a]	9.48 ± 0.62 [b]	7.70 ± 0.29 [a]

For the same milk, means with different superscript in the same row differ significantly ($p < 0.05$).

3.3. Somatic Cell Counts

Somatic cells were completely removed from both ovine and bovine permeates, and were concentrated in retentates (Table 1). One of the main purposes of applying MF is the 100% reduction in SCC. In the case of skimmed milk, this can be achieved using a MF membrane of 1.4 μm pore size [4,30], whereas, for raw whole bovine milk, membranes of an average pore size from 12 μm to 5 μm with permeate fluxes between 2000 L·m^{-2}·h^{-1} and 1460 L·m^{-2}·h^{-1} are needed to remove 93–100% of the SCC [5].

3.4. Microbial Counts

The populations of microorganisms are presented in Figure 2. It is obvious that MF affected the microbial groups in a different way, since the retention of specific bacteria depends on their cellular volume [31,32] and on their initials counts in the feed milk. In bovine permeate, total mesophilic microorganisms, coliforms, thermophilic lactococci and lactobacilli were reduced from 1.5 to 2.5 Log, while in ovine permeate, the microbial counts reduction was more efficient, e.g., total mesophilic microflora was reduced about 4 Log. Similar reduction of bacterial load has been reported for bovine permeate [13,17,31], while a reduction 2–3 Log has been reported for ovine permeate obtained with flux 200 L·m^{-2}·h^{-1} under pressure 0.6 bar at 40 °C [17]. It has been shown that by using a membrane of 1.4 μm pore size and fluxes of over 640 L·m^{-2}·h^{-1}, 99.7% of the bacteria can be removed from skim bovine milk [33]. Moreover, by cold MF (1.4 μm) at 6 °C, a method to inhibit bacteria growth in the system, an average of 3.4 Log reduction in vegetative bacteria can be achieved [9]. In the present study, the MF system and the applied conditions should meet the nominal reduction of bacteria counts 3–4 Log at flux 166 L·m^{-2}·h^{-1}. In the case of ovine milk, the flux was 105 ± 32 L·m^{-2}·h^{-1} and hence, this milk kind in combination with its higher protein content, which probably caused a thicker deposit layer on the membrane, retained more microorganisms than bovine milk.

Regarding sporeforming microorganisms, they were completely removed from ovine permeate, while in bovine permeate, they were reduced about 0.5–2 Log. The total retention of such microorganisms in the case of ovine milk was also attributed to a thicker protein layer formation on the membrane. In contrast, the insufficient retention of them in bovine retentate followed the trend of all other microbial groups and was attributed to the MF applied conditions in combination with the kinds of the present microorganisms. For example, *Lactobacillus casei*, a thermophilic microorganism, has cell width size 0.5–0.8 μm, whereas *B. cereus*, an aerobic sporeforming microorganism, is 1 μm [32]. Griep et al. [7], using cold MF (1.4 μm) for skim bovine milk, showed that *B. licheniformis* spores were reduced 2.17 log, while *Geobacillus sp.* spores were completely removed.

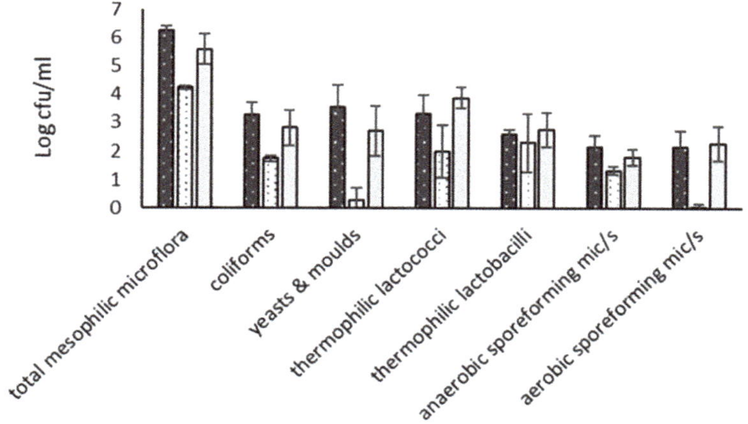

Figure 2. *Cont.*

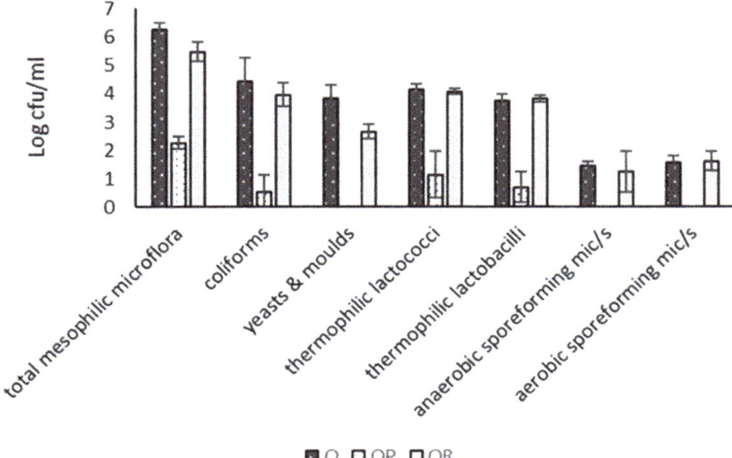

Figure 2. Effect of microfiltration on microbial counts (Log cfu.mL^{-1}) of partially defatted bovine (B) or ovine (O) milk, bovine permeate (BP), bovine retentate (BR) (mean ± SD, n = 3), ovine permeate (OP) and ovine retentate (OR) (mean ± SD, n = 5).

3.5. Alkaline Phosphatase Activity

The activity of Alkaline Phosphatase (ALP), an indigenous milk enzyme with technological importance in respect to the heat treatment of milk, was significantly reduced about 50–59% in bovine and ovine MF permeates (Table 4). This was because the γ-type of ALP is mostly found on the milk fat globule membrane [34] and, thus, its activity followed the partition of the milk fat after microfiltration. The ALP activity of permeate bovine was found to be as high as 407 µg phenol/mL, and it was similar to the value 445 µg phenol/mL reported for bovine skim milk, i.e., 0.05% fat [35]. The same researchers have reported the ALP of ovine skim milk as 4382 µg phenol/mL.

Table 4. Effect of microfiltration on alkaline phosphatase (ALP) activity (µg phenol/mL) and protease activities (area × 10^6 of peaks of RP-HPLC) in the whey fraction of partially defatted bovine (B) or ovine (O) milk, bovine permeate (BP), bovine retentate (BR) (mean ± SD, n = 3), ovine permeate (OP) and ovine retentate (OR) (mean ± SD, n = 5).

	Bovine			Ovine		
	B	**BP**	**BR**	**O**	**OP**	**OR**
ALP	810 ± 127 [a]	407 ± 111 [b]	964 ± 217 [a]	4728 ± 958 [a]	2832 ± 757 [b]	4132 ± 1009 [a]
Proteases in whey						
Cathepsin D-like product	0.99 ± 0.08 [a]	0.97 ± 0.08 [a]	1.24 ± 0.03 [b]	0.39 ± 0.06	0.34 ± 0.05	0.36 ± 0.01
Peak 1	1.04 ± 0.08 [a]	1.06 ± 0.14 [a]	0.64 ± 0.11 [b]	2.02 ± 0.25	1.93 ± 0.15	1.81 ± 0.18
Peak 2	0.21 ± 0.04	0.19 ± 0.07	0.21 ± 0.04	0.07 ± 0.01 [a]	0.16 ± 0.02 [b]	0.19 ± 0.02 [b]
Peak 3	0.11 ± 0.01 [a]	0.04 ± 0.01 [b]	0.04 ± 0.01 [b]	0.14 ± 0.05	0.13 ± 0.06	0.25 ± 0.03

For the same milk, means with different superscript in the same row differ significantly ($p < 0.05$).

3.6. Cathepsin D Activity

Cathepsin D, a lysosomal aspartic proteinase, is found in somatic cells and acts on κ-, α_{s1}- and β-casein similarly to how it acts on chymosin [36]. The assay used for its determination was based on the presence of peak P (Figure 3), which is the product that resulted from the action of cathepsin D on the substrate. The quantitative determination (Table 4) was based on the chromatographic area of this peak. Peak 1 is another aspartic proteinase, whereas peaks 2 and 3 are possible cysteine- proteinases [24]. Cathepsin D was determined in the acid whey fraction of the feed milks, MF permeates and retentates, and there was no significant difference among them, since whey is easily filtered through a membrane of 1.4 μm pore size. However, it is noteworthy that although somatic cells were completely removed from permeates, the acid proteinases associated with them were present. It seems that the shear forces induced by circulation removes cathepsin from somatic cells and, thus, it remains in the permeate. The higher flux in the case of bovine milk might cause higher dissociation from somatic cells and, therefore, the product resulted from cathepsin activity was higher than that of ovine milk.

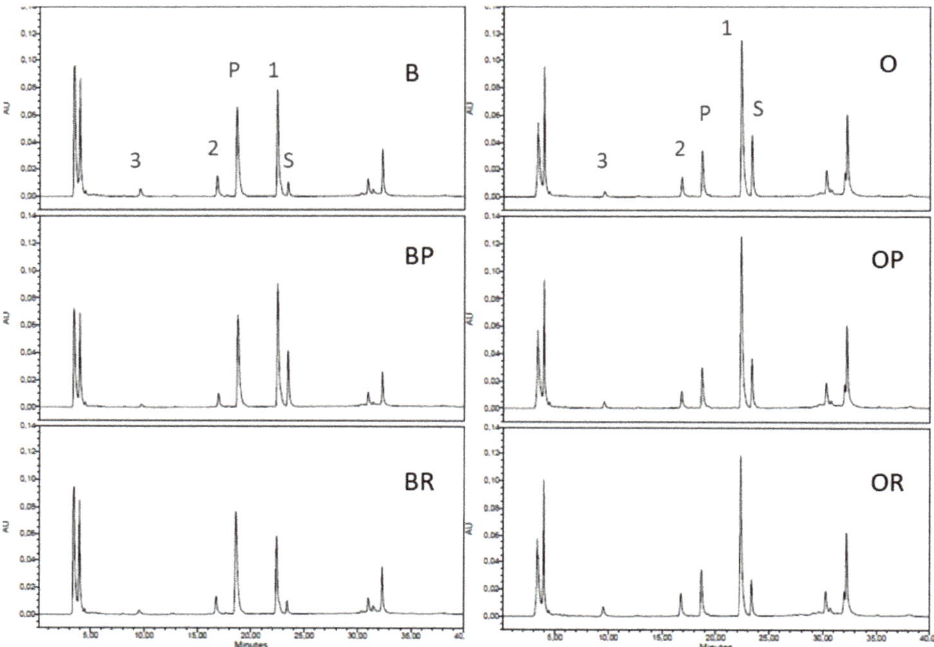

Figure 3. Proteinase activities in the acid whey fractions of feed bovine milk (B), bovine permeate (BP) and bovine retentate (BR), or feed ovine milk (O), ovine permeate (OP) and ovine retentate (OR), after incubation with the substrate Pro-Thr-Glu-Phe-[p-nitro-Phe]-Arg-Leu at pH 3.2 for 12 h at 37 °C. S: residual substrate; P: cathepsin D-like product; peaks 1, 2 and 3: other proteinases.

3.7. Renneting Behavior

Regarding the renneting behavior, there was not a significant difference among the feed, the permeate and retentate for both milks as far as the milk clotting time was concerned (Table 5). Milk clotting time depends mainly on the ratio enzyme/substrate, milk pH, temperature and calcium ions. Although in the case of ovine milk, calcium and phosphorous contents in the permeate were significantly lower than in feed milk or MF retentate, and the clotting time was not significantly affected. In addition, the firmness of the curd made from ovine permeate did not differ significantly from the curd made from ovine feed milk, although the casein content of milk has a significant influence on its maximum firmness [26]. In contrast, the firmness of curd made from bovine permeate was significantly lower than the curd made from the feed milk.

Table 5. Effect of microfiltration on milk clotting time (r) and curd firmness (A_{30}) of partially defatted bovine (B) or ovine (O) milk, bovine permeate (BP), bovine retentate (BR) (mean ± SD, n = 3), ovine permeate (OP) and ovine retentate (OR) (mean ± SD, n = 5).

	Bovine			Ovine		
	B	**BP**	**BR**	**O**	**OP**	**OR**
r (min)	20.71 ± 1.13	22.04 ± 2.41	21.25 ± 2.08	14.15 ± 2.44	14.41 ± 2.25	14.34 ± 2.38
A_{30} (mm)	19.12 ± 2.93 [a]	12.58 ± 3.22 [b]	17.81 ± 2.26 [a,b]	42.71 ± 3.73 [a]	39.70 ± 4.35 [a,b]	46.29 ± 4.55 [b]

For the same milk, means with different superscript in the same row differ significantly ($p < 0.05$).

4. Conclusions

From the obtained results it appears that the application of crossflow MF using a membrane of 1.4 μm pore size, at 50 °C and TMP 1.5 bar, influences the ovine milk (0.4% fat) in a similar way to the bovine milk (0.3% fat). More specifically, the applied crossflow MF improved the microbial quality, but significantly reduced ($p < 0.05$) the protein content and, consequently, the total solids content of both the ovine and bovine permeate. In addition, crossflow MF influenced the distribution of casein fractions between permeate and retentate with αs1- and β-CN retention being more pronounced in the case of ovine milk. The activity of indigenous enzyme ALP followed the allocation of the fat content, while the activity of cathepsin D was not influenced. Regarding cheesemaking properties, the firmness of the curd made from the ovine feed milk did not differ from the curd made from the ovine permeate. It is concluded that crossflow MF under the studied conditions can be used as pre-treatment to improve the microbial quality of ovine milk prior to cheesemaking. However, further study is needed to optimize the conditions of crossflow MF processing for this kind of milk.

Author Contributions: Conceptualization, E.M.; Data curation, G.P. and E.M.; Formal analysis, G.P. and C.P.; Methodology, E.M. and G.M.; Project administration, E.M.; Writing—review & editing, E.M. All authors have read and agreed to the published version of the manuscript.

Funding: This research received no external funding.

Acknowledgments: This study is dedicated to the memory of I. Kandarakis who introduced the membrane technology in our lab. Moreover, the authors would like to thank T. Paschos for the technical support on microfiltration plant and the Dairy Industry DELTA S.A. for milk analyses on Milkoscan and Fossomatic.

Conflicts of Interest: The authors declare no conflict of interest.

References

1. Rosenberg, M. Current and future application for membrane processes in the dairy industry. *Trends Food Sci. Technol.* **1995**, *6*, 12–19. [CrossRef]
2. Pouliot, Y. Membrane processes in dairy technology—From a simple idea to worldwide panacea. *Int. Dairy J.* **2008**, *18*, 735–740. [CrossRef]
3. Merin, U.; Dafin, G. Cross flow microfiltration in the dairy industry: State-of-the-art. *Le Lait* **1990**, *70*, 281–291. [CrossRef]
4. Saboya, L.V.; Maubois, J.L. Current developments of microfiltration technology in the dairy industry. *Le Lait* **2000**, *80*, 541–553.
5. Maubois, J.L.; Schuck, P. Membrane technologies for the fractionation of dairy components. *Bul. Int. Dairy Federation.* **2005**, *400*, 2–7.
6. Guerra, A.; Jonsson, G.; Rasmussen, A.; Nielsen, E.W.; Edelsten, D. Low cross flow velocity microfiltration of skim milk for removal of bacterial spores. *Int. Dairy J.* **1998**, *7*, 849–861. [CrossRef]
7. Griep, E.R.; Cheng, Y.; Moraru, C. Efficient removal of spores from skim milk using cold microfiltration: Spore size and surface property considerations. *J. Dairy Sci.* **2018**, *101*, 9703–9713. [CrossRef]
8. Hoffmann, W.; Kiesner, C.; Clawin-Rädecker, I.; Martin, D.; Einhoff, K.; Lorenzen, C.P.; Meisel, H.; Hammer, P.; Suhren, G.; Teufel, P. Processing of extended shelf life milk using microfiltration. *Int. J. Dairy Technol.* **2006**, *59*, 229–235. [CrossRef]

9.   Wang, D.; Fritsch, J.; Moraru, C. Shelf life and quality of skim milk processed by cold microfiltration with a 1.4-μm pore size membrane, with or without heat treatment. *J. Dairy Sci.* **2019**, *102*, 8798–8806. [CrossRef]

10. Goudédranche, H.; Fauquant, J.; Maubois, J.-L. Fractionation of globular milk fat by membrane microfiltration. *Le Lait* **2000**, *80*, 93–98. [CrossRef]

11. Lawrence, N.D.; Kentish, S.E.; O'Connor, A.J.; Barber, A.R.; Stevens, G.W. Microfiltration of skim milk using polymeric membranes for casein concentrate manufacture. *Sep. Purif. Technol.* **2008**, *60*, 237–244. [CrossRef]

12. Soodam, K.; Guinee, G. The case for milk protein standardisation using membrane filtration for improving cheese consistency and quality. *Int. J. Dairy Technol.* **2018**, *71*, 277–291. [CrossRef]

13. Pafylias, I.; Cheryan, M.; Mehaia, M.A.; Saglam, N. Microfiltration of milk with ceramic membranes. *Food Res. Int.* **1996**, *29*, 141–146. [CrossRef]

14. Elwell, M.W.; Barbano, D.M. Use of microfiltration to improve fluid milk quality. *J. Dairy Sci.* **2006**, *89* (Suppl. E), E20–E30. [CrossRef]

15. European Union 2006. Commission Regulation (EC) No 1662/2006 of 6 November 2006 amending Regulation (EC) No 853/2004 of the European Parliament and of the Council laying down specific hygiene rules for food of animal origin. *Off. J. Eur. Union L* **2006**, *320*, 1–10.

16. Beolchini, F.; Veglio, F.; Barba, D. Microfiltration of Bovine and Ovine Milk for the Reduction of microbial Content in tubular membrane: A preliminary investigation. *Desalination* **2004**, *161*, 251–258. [CrossRef]

17. Beolchini, F.; Cimini, S.; Mosca, L.; Veglio, F.; Barba, D. Microfiltration of Bovine and Ovine Milk for the Reduction of Microbial Content: Effect of Some Operating Conditions on Permeate Flux and Microbial Reduction. *Sep. Sci. Technol.* **2005**, *40*, 757–772. [CrossRef]

18. AOAC. *Official Methods of Analysis*, 12th ed.; Association of Official Analytical Chemists: Washington, DC, USA, 1975; p. 254.

19. IDF Standard 42: 2006. *Milk—Determination of Total Phosphorous Content—Method Using Molecular Absorption Spectrometry*; International Dairy Federation: Brussels, Belgium, 2006.

20. IDF Standard 119: 2007. *Milk and Milk Products—Determination of Calcium, Sodium, Potassium and Magnesium Content-Atomic Absorption Spectrometric Method*; International Dairy Federation: Brussels, Belgium, 2007.

21. Moatsou, G.; Moschopoulou, E.; Molle, D.; Gagnaire, V.; Kandarakis, I.; Leonil, J. Comparative study of the protein fraction of goat milk from the Indigenous Greek breed and from international breeds. *Food Chem.* **2008**, *106*, 509–520. [CrossRef]

22. O'Driscoll, B.M.; Rattray, F.P.; McSweeney, P.L.H.; Kelly, A.L. Protease activities in raw milk determined using a synthetic heptapeptide substrate. *J. Food Sci.* **1999**, *64*, 606–611. [CrossRef]

23. Hurley, M.J.; Larsen, L.B.; Kelly, A.L.; McSweeney, P.L.H. Cathepsin D activity in quarg. *Int. Dairy J.* **2000**, *10*, 453–458. [CrossRef]

24. Moatsou, G.; Katsaros, G.; Bakopanos, C.; Kandarakis, I.; Taoukis, P.; Politis, I. Effect of high-pressure treatment at various temperatures on activity of indigenous proteolytic enzymes and denaturation of whey proteins in ovine milk. *Int. Dairy J.* **2008**, *18*, 1119–1125. [CrossRef]

25. IDF Standard 63: 2009. *Milk—Determination of Alkaline Phosphatase*; International Dairy Federation: Brussels, Belgium, 2009.

26. Park, Y.W.; Juarez, M.; Ramos, M.; Haenlein, G.F.W. Physico-chemical characteristics of goat and sheep. *Small Rum. Res.* **2007**, *68*, 88–113. [CrossRef]

27. Piry, A.; Kuhnl, W.; Grein, T.; Tolkach, A.; Ripperger, S.; Kulozik, U. Length dependency of flux and protein permeation in crossflow microfiltration of skimmed milk. *J. Membr. Sci.* **2008**, *325*, 887–894. [CrossRef]

28. Tziboula, A.; Steele, W.; West, I.; Muir, D.D. Microfiltration of milk with ceramic membranes: Influence on casein composition and heat stability. *Milchwissensichaft* **1998**, *53*, 8–11.

29. Heidebrecht, H.J.; Kulozik, U. Fractionation of casein micelles and minor proteins by microfiltration in diafiltration mode. Study of the transmission and yield of the immunoglobulins IgG, IgA and IgM. *Int. Dairy J.* **2019**, *93*, 1–10. [CrossRef]

30. Te Giffel, M.C.; van der Horst, H.C. Comparison between bactofugation and microfiltration regarding efficiency of somatic cell and bacteria removal. *Bul. Int. Dairy Federation.* **2004**, *389*, 49–53.

31. Trouve, E.; Maubois, J.L.; Piot, M.; Madec, M.N.; Fauquant, J.; Rouault, A.; Tabard, J.; Brinkman, G. Retention de differentes especes microbiennes lors de l'epuration du lait par microfiltration en flux tangential. *Le Lait* **1991**, *72*, 327–332.

Foods **2020**, *9*, 284

32. Fernandez Garcia, L.; Alvarez Blanco, S.; Riera Rondriguez, A. Microfiltration applied to dairy streams: Removal of bacteria. *J. Sci. Food Agric.* **2013**, *93*, 187–196. [CrossRef]
33. Malmberg, M.; Holm, S. Low bacteria skim milk by microfiltration. *North Eur. Food Dairy J.* **1988**, *54*, 75–78.
34. Shakeel-ur-Rehman, S.; Fleming, C.M.; Farkye, N.Y.; Fox, P.F. Indigenous phosphatases in milk. In *Advanced Dairy Chemistry Volume I —Proteins*; Fox, P.F., McSweeney, P.L.H., Eds.; Kluwer Academic-Plenum Publishers: New York, NY, USA, 2003; pp. 523–543.
35. Dumitraşcu, L.; Stănciuc, N.; Stanciu, S.; Râpeanu, G. Inactivation Kinetics of Alkaline Phosphatase from Different Species of Milk Using Quinolyl Phosphate as a Substrate. *Food Sci. Biotechnol.* **2014**, *23*, 1773–1778. [CrossRef]
36. Moatsou, G. Indigenous enzymatic activities in ovine and caprine milks. *Int. J. Dairy Technol.* **2010**, *63*, 16–31. [CrossRef]

Article

Hypoallergenic and Low-Protein Ready-to-Feed (RTF) Infant Formula by High Pressure Pasteurization: A Novel Product

Md Abdul Wazed and Mohammed Farid *

Department of Chemical and Materials Engineering, University of Auckland, Private Bag 92019, Auckland 1142, New Zealand; mwaz610@aucklanduni.ac.nz
* Correspondence: m.farid@auckland.ac.nz; Tel.: +64-9-9234807

Received: 20 July 2019; Accepted: 10 September 2019; Published: 12 September 2019

Abstract: Infant milk formula (IMF) is designed to mimic the composition of human milk (9–11 g protein/L); however, the standard protein content of IMF (15 g/L) is still a matter of controversy. In contrast to breastfed infants, excessive protein in IMF is associated with overweight and symptoms of metabolic syndrome in formula-fed infants. Moreover, the beta-lactoglobulin (β-Lg) content in cow milk is 3–4 g/L, whereas it is not present in human milk. It is considered to be a major reason for cow milk allergy in infants. In this respect, to modify protein composition, increasing the ratio of alpha-lactalbumin (α-Lac) to β-Lg would be a pragmatic approach to develop a hypoallergenic IMF with low protein content. Such a formula would ensure the necessary balance of essential amino acids, as 123 and 162 amino acid residues are available in α-Lac and β-Lg, respectively. Hence, in this study, a pasteurized form of hypoallergenic and low-protein ready-to-feed (RTF) formula, a new product, is developed to retain heat-sensitive bioactives and other components. Therefore, the effects of high pressure processing (HPP) under 300–600 MPa at approximately 20–40 °C and HTST pasteurization (72 °C for 15 and 30 s) were investigated and compared. The highest ratio of α-Lac to β-Lg was achieved after HPP (600 MPa for 5 min applied at 40.4 °C), which potentially explains the synergistic effect of HPP and heat on substantial denaturation of β-Lg, with significant retention of α-Lac in reconstituted IMF. *Industrial relevance*: This investigation showed the potential production of a pasteurized RTF formula, a niche product, with a reduced amount of allergenic β-Lg.

Keywords: alpha-lactalbumin (α-Lac); beta-lactoglobulin (β-Lg); high pressure processing (HPP); pasteurization; ready-to-feed (RTF) infant formula

---

1. Introduction

Infant milk formula (IMF) is intended to serve as a functional substitute for infants under 12 months of age. The World Health Organization (WHO) recommends absolute breastfeeding of infants for the first 6 months of age [1], whereas the American Society of Paediatrics suggests the same for at least 12 months [2]. However, only 38% of infants are being breastfed globally [3], which indicates the common use of IMF, a $41 billion USD market [4]. IMF is available in three forms—powdered, liquid concentrate, and liquid ready-to-feed (RTF). Among them, RTF, the most convenient form, is currently manufactured as a sterilized product to ensure safety using UHT (Ultra High Temperature, 135–145 °C for a short time). However, UHT causes the nutritional profile of RTF formula to deteriorate, especially vitamin A, B, and D, along with protein denaturation, which requires supplementation of micronutrients (e.g., vitamins and minerals) [5]. Although every possible effort is being made to bring IMF closer to human milk (HM), there is still a gap between them nutritionally, which governs neurological, physiological, and immunological growth and development of infants [6–8]. Such a difference could be explained by the nutritional balance that remains fixed in an IMF, while it varies in HM throughout the lactation period and even between individuals [9].

Moreover, beta-lactoglobulin (β-Lg) represents about 50% of total whey as a major whey protein in cow milk. It is also considered a major allergen to infants despite having numerous functional and nutritional roles in adult human health. However, interestingly, this β-Lg is absent in HM. To address the cow's milk allergy (CMA), researchers and manufacturers introduced partially hydrolyzed formula (pHF) and extensively hydrolyzed formula (eHF), which are also recommended by paediatricians to reduce early allergy manifestation [10]. In pHF and eHF, enzymatic hydrolysis of proteins with digestive enzymes reduces the allergic properties by breaking them into small peptides and free amino acids, which are not allergenic [11]. However, these hydrolyzed formulas often exhibit bitter taste, poor flavour, reduced lipid emulsifiability, and elevated osmolality, which limit their application in general IMF [12].

On the other hand, alpha lactalbumin (α-Lac) is a bioactive protein present in all mammal milk, which is regarded as a component of lactose synthesis with antimicrobial, prebiotic, and Ca-binding capacity [13]. The α-lac content in human milk is 3–4 g/L, while it is only 1 g/L in mature cow milk. It contains a high level of different essential amino acids like tryptophan, lysine, and cysteine, and hence, fortification of IMF with α-Lac is recommended [13,14].

The standard protein content of infant formula is still a matter of controversy since formula production aims to mimic HM [14]. HM contains 9–11 g/L protein [15], while conventional infant formula provides 15 g/L [16]. Burgeoning demand for low-protein infant formula, especially in Asia, resulted mainly from paediatric obesity [17]. Moreover, excess protein also induces unnecessary strain on immature metabolic organs [18,19]. Thus, the possible alternative could be reducing the protein content and adding free amino acid into infant formula, which would be unphysiological since metabolic consequences of free amino acids are mostly unknown [14]. Thereupon, a logical approach would be to modify the protein profile of formula to make it closer to HM by increasing the ratio of α-Lac to β-Lg, which will yield a formula that has lower total protein, but retains the necessary balance of essential amino acids [14,16,20,21]. Moreover, this would also be an imperative strategy because complete removal of allergenic β-Lg would deteriorate the function of milk proteins since it is necessary for whey proteins to be associated with casein micelles [22]. High pressure processing (HPP) defragments casein micelles into smaller particles and splits them into more soluble components like αs1-, αs2-, β-, and k-caseins [23–25]. More than a 50% reduction in the size of casein micelle was observed after HPP at >300 MPa at 40 °C [24]. In powdered IMF, reducing the protein content to as low as 9.77 g protein/100 g did not cause a significant impact on physical stability and shelf-life [26]. However, in RTF liquid IMF, a higher ratio of α-Lac to β-Lg reduces the viscosity and induces rapid sedimentation during storage, as observed in UHT-treated products; and hence, the addition of thickener and stabilizer is suggested to manage the viscosity and sedimentation, respectively [27–29]. Furthermore, Crowley et al. [27] reported increased α-Lac reduced heat-induced coagulation in a model whey protein-dominant IMF. They further observed lower protein–protein interactions in the model IMF due to the fortification of IMF with α-Lac. Additionally, Kamarei [30] patented his invention in manufacturing refrigeration-shelf-stable pasteurized IMF with required quantities of nutrients. In this patent, he reported that UHT-treated RTF formula provides a different and unknown amount of degradable micro nutrients due to the high-heat treatment and subsequent storage, which also affects the nutritional value and sensory attributes. These outcomes evidence the possibility of developing an IMF with a higher ratio of α-Lac to β-Lg.

Previously, Huppertz et al. [31] described the mechanism of HPP-induced denaturation of α-Lac and β-Lg at 200–800 MPa/20 °C in whole milk and reported α-Lac as a pressure-resistant protein, unlike β-Lg. Furthermore, Mazri et al. [32] also investigated the denaturation kinetics of these two bioactive proteins in skim milk under HPP of 450–700 MPa at 20 °C and confirmed the baroresistance of α-Lac in comparison to β-Lg. Recently, HPP has been recommended to preserve human milk, due to its efficient inactivation of microbial pathogens, along with the retention of unique components [33].

Wesolowska et al. [34] also reviewed the effect of HPP to pasteurize human milk and referred to HPP as superior to the conventional holder pasteurization (63 °C/30 min) in maintaining bioactivity of protein components. To the best of our knowledge, the synergistic effect of HPP and heat to achieve a higher ratio of α-Lac to β-Lg has not been explored in IMF, which fundamentally pioneered this work.

Therefore, the aim of this work was to achieve a higher ratio of α-Lac to β-Lg through investigating their retention after HPP at different pressure–temperature–time combinations in reconstituted IMF, fortified with α-Lac. For comparison, we also performed high-temperature short-time (HTST) pasteurization at 72 °C for 15 and 30 s, and thereafter measured the concentration of α-Lac and β-Lg. The combined effect of pressure and temperature on the kinetics of denaturation of both proteins was also analyzed.

2. Materials and Methods

2.1. α-Lac-Added Reconstituted IMF Preparation

Bovine α-Lac powder (native form) was generously donated by Davisco Foods International, Le Sueur, MN, USA. The protein content and α-Lac level in the powder were more than 95% and 90%, respectively. Spray-dried IMF powder (stage 1, 1.4 g protein/100 mL, whey protein to casein ratio 60:40) was reconstituted into cooled, boiled milli-Q water, as per the instructions of the manufacturer. Finally, 100 mg of α-Lac was added into 100 mL of reconstituted IMF, and a magnetic stirrer was used for gentle mixing.

2.2. HTST Treatment

HTST treatments at 72 °C for 15 and 30 s were performed in duplicate, using a sample (15 mL) in a copper tube, then immersing it in a water bath (72.2 °C). A thermocouple (K-type) was inserted at the geometric centre of the copper tube (29 cm length, 9.6 mm outer diameter, and 8.0 mm inner diameter) to record the temperature–time profile using a data logger. The sample temperature increased sharply within 30 s to achieve the steady-state temperature. The sample tube was immediately transferred to an ice bath after treatment. It is to be noted that the determination of α-Lac and β-Lg content following ELISA (Section 2.4) requires a large amount of sample (15 mL). For this reason, this study could not follow the capillary method to perform HTST treatments.

2.3. HPP Treatment

A QFP 2L-700 HPP unit (Avure Technologies, Columbus, OH, USA) was used to perform the combined pressure–heat treatments. The equipment has a 2 L cylindrical stainless steel pressure treatment chamber with two internal thermocouples for monitoring the temperature. The system also includes a heating system, water circulation, and a pumping system, along with a computer-operated control system. Distilled water was used as a pressure transmission fluid, and isostatic pressure transmission resulted in a uniform pressure in all directions [35].

In HPP treatments, vacuum-packaged 15 mL samples were placed into the treatment chamber and immediately cooled in an ice water bath after treatment. HPP of 300 MPa (10 and 20 min), 400 MPa (10 and 20 min), 500 MPa (5 and 10 min), and 600 MPa (1 and 5 min) were applied at different temperatures, ranging from ambient temperature to ~40 °C. The treatment time refers to the duration of steady-state pressure conditions, as programmed in the operating system. The temperature increase was about 2.3 °C/100 MPa, due to the adiabatic heating during pressurization, and the decompression time was <20 s in all treatments.

However, similar to Evelyn and Silva [36], the temperature of the HPP chamber dropped steadily during the steady-state pressure phase of the HPP cycle, due to cooling followed by a rapid drop in temperature during decompression (Figure 1).

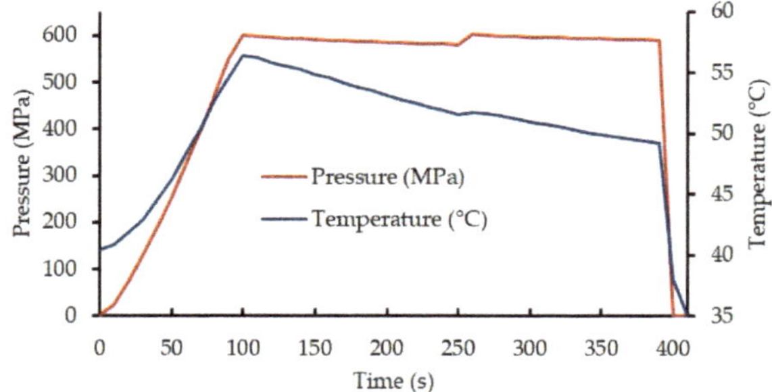

Figure 1. Temperature-pressure profile in alpha-lactalbumin (α-Lac)-added reconstituted infant milk formula (IMF): High pressure processing (HPP) applied at 40.4 °C/600 MPa for 5 min.

Transient temperature change significantly impacts all chemical and biological reactions, due to its exponential effect on them. The average temperature during the steady-state pressure phase does not represent the accurate representation of the treatment temperature. For instance, as shown in Figure 1, the pressure come-up time (100 s) prior to achieving the steady-state pressure condition is relatively long and must be accounted for. Considering this, Farid and Alkhafaji [37] provided an integrated value of processing temperature, named "effective treatment temperature (T_{eff})", to represent the entire treatment period (Equation (1)).

$$T_{\text{eff}} = \frac{-E/R}{\ln(\sum_0^n e^{-E/RT}/n - 1)} \tag{1}$$

where T (K) is the absolute temperature recorded at a given time interval (t), E is the activation energy (kJ/mol) of the specific biological or chemical reaction, R is the gas constant (8.314 J/mol/K), and n is the number of temperature recording points measured at equal time intervals.

From now on, the temperature in HPP treatments referred to in this paper will be the T_{eff}.

2.4. Sample Preparation for α-Lac and β-Lg Determination

Defatted whey supernatant was prepared from both HTST- and HPP-treated samples by centrifuging them at 9000 g for 15 min at 4 °C. Then, the whey portion was obtained by adjusting the pH to 4.6 using 8 M acetic acid, followed by centrifugation at the same conditions. The pH of the supernatant was readjusted to 6.8 using 3 M NaOH.

2.5. Determination of α-Lac and β-Lg Content

An enzyme-linked immunosorbent assay (ELISA) method was followed to measure the concentration of α-Lac and β-Lg. The sandwich ELISA was performed using bovine α-Lac and β-Lg Quantitation Kits (catalogue E10-128 and E10-125, respectively) and an ELISA Starter Accessory Kit (catalogue E103), purchased from Bethyl Laboratories, TX, USA. The ELISA plates were coated, and supplied bovine α-Lac and β-Lg were diluted to obtain standard curves, following the manufacturer's protocols. A multimode plate reader with associated software (Perkin Elmer's EnSpire Multimode Plate reader, Waltham, Massachusetts, USA) was used to read the absorbance of ELISA plates at 450 nm. The unknown concentration of α-Lac and β-Lg of treated samples was obtained in duplicate from the standard curves. Retention of α-Lac and β-Lg was calculated as a percentage of α-Lac and β-Lg in untreated samples, respectively, as given in Equation (2).

$$\text{Retention} = \frac{c_t}{c_o} \times 100, \tag{2}$$

where c_t and c_o represent the concentrations of α-Lac or β-Lg after and before treatment, respectively.

2.6. Kinetics of Protein Denaturation

Activation energy (Ea) for α-Lac and β-Lg denaturation was calculated using the results obtained from HPP-treated samples. The denaturation of α-Lac and β-Lg with time after HPP treatment can be described by the general rate equation

$$-\frac{dc}{dt} = kc^n, \tag{3}$$

where $-\frac{dc}{dt}$ is the rate of denaturation, k is the rate constant, c is the concentration of α-Lac or β-Lg, and n is the order of reaction.

It has been reported that HPP denaturation of α-Lac follows first-order kinetics in whole milk [38]. For first-order kinetics ($n = 1$), the integration of Equation (3) gives

$$\ln(c_t/c_o) = -kt. \tag{4}$$

The semi-logarithmic plot of Equation (4) gives a straight line with high coefficients of correlation (r^2), and the value of the ordinate intercept b (time, $t = 0$) appears close to zero. The slopes of the lines obtained correspond to the rate constant (k).

Furthermore, the reaction kinetics for β-Lg in HPP was also reported as a second-order reaction by Anema et al. [39] in skim milk and Hinrichs et al. [40] in whey proteins. For non-first order kinetics ($n \neq 1$), the integration of Equation (3) gives

$$(c_t/c_o)^{1-n} = 1 + (n-1)kt. \tag{5}$$

The graphical representation of Equation (5) yields straight lines, and the ordinate intercept b (time, $t = 0$) should be 1 if the treatment follows the estimated reaction order. The rate constant (k) is obtained from the slope of the lines. The Arrhenius equation relates the treatment temperature and the rate constant of a denaturation process as given in Equation (6):

$$k = Ae^{-E_a/RT}, \tag{6}$$

where A is the pre-exponential factor, E_a is the activation energy, R is the universal gas constant, and T is the absolute temperature.

By taking the logarithms of both sides, Equation (6) gives a linearized form as

$$\ln k = (-Ea/R) \times (1/T) + \ln A. \tag{7}$$

The graphical representation of Equation (7) (lnk vs. $1/T$) determines the effect of temperature on the rate of constant (k). The gradient of Equation (7) is equal to $-Ea/R$, and thus the activation energy (Ea) is calculated.

2.7. Statistical Analysis

Statistical analysis of data, including mean and standard deviation for replicates, was performed using Microsoft Excel 2013 (Microsoft Inc., Redmond, Washington, USA). The level of significance was set at $p = 0.05$.

3. Results and Discussion

3.1. Calculation of Effective Treatment Temperature (T_{eff})

In this study, T_{eff} for HPP treatments on α-Lac and β-Lg was calculated using Equation (1) and tabulated in Table 1. In all cases, T_{eff} was significantly lower than the maximum temperature in HPP (T_{max}), which further establishes the application of the concept of T_{eff} in denaturation studies.

Table 1. HPP conditions at different temperatures and the corresponding parameters.

Pressure	300 MPa			400 MPa			500 MPa			600 MPa		
Pressure come up time (s)	60	60	60	80	80	80	90	90	90	100	100	100
Initial temperature (°C)	14.8	30.4	41.2	14.7	29.9	40.2	14.3	29.9	40.7	21.2	30.3	40.4
Maximum temperature, T_{max}(°C)	21.6	38.5	51.2	23.9	40.2	50.8	27.2	42.3	54.6	35.9	45.1	56.4
T_{eff} (°C)	18.1	36.2	46.5	21.2	36.6	47	23.8	38.1	49.2	31.9	41	51.7

3.2. Kinetics of α-Lac and β-Lg Denaturation during HPP Treatments

Activation energy (*Ea*) for α-Lac and β-Lg denaturation was calculated using the results obtained from HPP treated samples following Section 2.6. The reaction orders for HPP-induced denaturation of α-Lac and β-Lg were determined to compare the rate constants (*k*) at different temperatures and pressures, and to calculate the *Ea*. Experimental data obtained in this work and their graphical representation in Figures 2 and 3 yielded the reaction order (*n*) as 1 and 2 for α-Lac and β-Lg, respectively. These orders consistently produced reasonably straight lines with good correlation of coefficients ($r^2 > 0.98$) and agree well with previous studies [38–40].

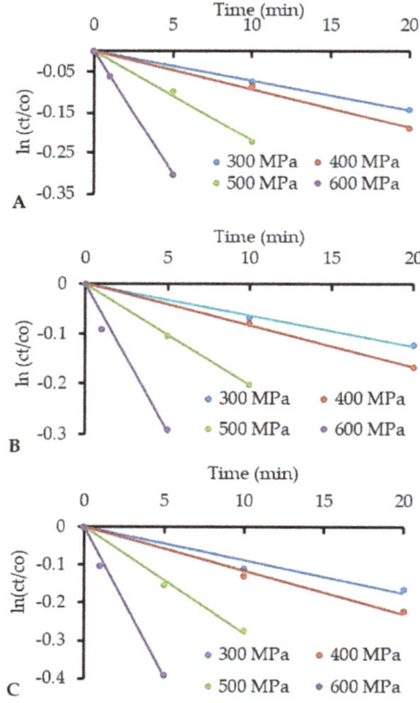

Figure 2. High-pressure denaturation of alpha-lactalbumin (α-Lac), HPP applied at ~20 °C (**A**), ~30 °C (**B**), and ~40 °C (**C**). Temperature at all HPP conditions (T_{eff}) is presented in Table 1. The concentration of α-Lac is expressed as c_t/c_o, where c_t = α-Lac concentration after HPP and c_o = initial α-Lac concentration before HPP.

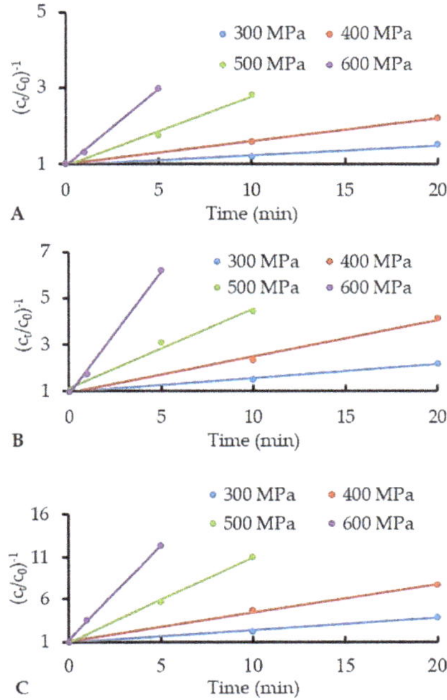

Figure 3. High-pressure denaturation of beta-lactoglobulin (β-Lg), HPP applied at ~20 °C (**A**), ~30 °C (**B**), and ~40 °C (**C**). Temperature at all HPP conditions (T_{eff}) is presented in Table 1. The concentration of β-Lg is expressed as c_t/c_o, where c_t = β-Lg concentration after HPP and c_o = initial β-Lg concentration before HPP.

Furthermore, Figures 4 and 5 represent the effect of HPP on the rate constant (k) for the denaturation of α-Lac and β-Lg, respectively. E_a was calculated from the gradients of respective lines from Figure 4 for α-Lac and from Figure 5 for β-Lg. E_a indicates the energy barrier that a protein is required to overcome to take part in a reaction. Values of E_a during HPP are presented in Table 2 where we observed a distinctive higher E_a in β-Lg denaturation compared to α-Lac. These results correspond well with the previous studies investigated by Huppertz et al. [31] for whole milk and Mazri et al. [32] for skim milk.

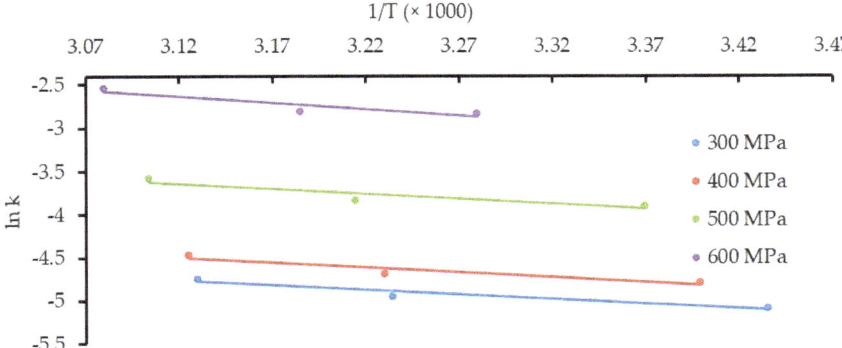

Figure 4. Effect of HPP on the rate of constant (k) for denaturation of α-Lac. Temperature at all HPP conditions (T_{eff}) is presented in Table 1.

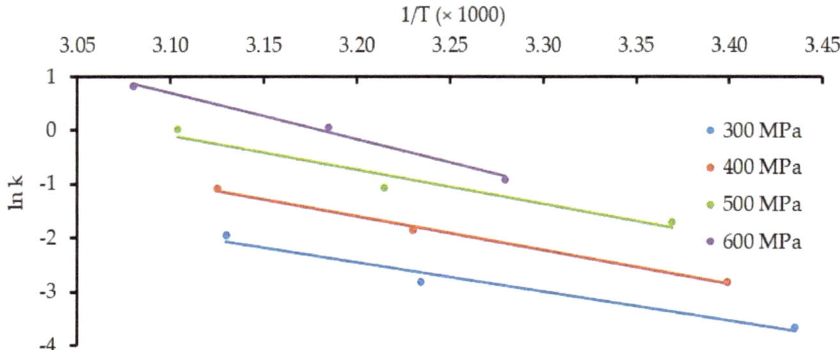

Figure 5. Effect of HPP on the rate of constant (k) for denaturation of β-Lg. Temperature at all HPP conditions (T_{eff}) is presented in Table 1.

Table 2. Activation energy (E_a) after HPP of 300–600 MPa.

HPP	Activation Energy, E_a (kJ/mol)	
	α-Lac	β-Lg
300 MPa	8.74	44.8
400 MPa	9.44	51.3
500 MPa	9.65	52.5
600 MPa	12.13	71.6

However, E_a values obtained in this work for β-Lg denaturation differ to some extent from those reported earlier by Anema et al. [39] from HPP treatments of 200–600 MPa at 10–40 °C up to 60 min. They reported E_a as 103.8 and 114.35 kJ/mol for 500 and 600 MPa, respectively; twice higher than the reported values of this work.

IMF is a complex food containing a variety of ingredients. Therefore, this difference in E_a could be attributed to the dissimilarity in treatment media, treatment duration, and the estimation of treatment temperature, since this study considered the temperature attained during the pressure come-up time.

3.3. Ratio of α-Lac to β-Lg

In this study, the retention of α-Lac and β-Lg after HTST and HPP was measured by ELISA and compared thereafter (Figure 6). Before treatment, the concentration of α-Lac and β-Lg in reconstituted IMF was 1.04 and 6.2 mg/mL, respectively. Conventional HTST (72 °C for 15 and 30 s) retained 78% and 70% of α-Lac and β-Lg respectively, which corroborates the results reported previously [27,41]. However, in contrast, the degree of denaturation of β-Lg was more pronounced than α-Lac, as observed with HPP combinations of increased pressure, temperature, and time (Figure 6). This trend is in agreement with those obtained from HPP in skim milk [31,32] and in a protein solution [42].

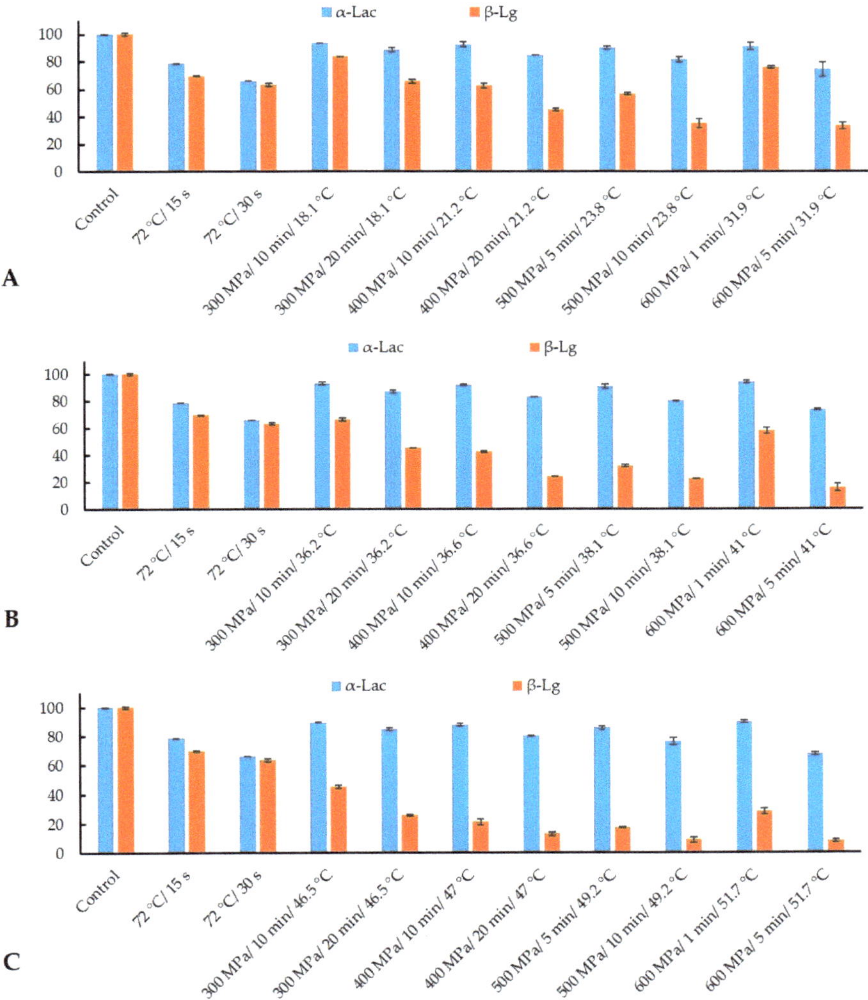

Figure 6. Retention of α-Lac and β-Lg (%) after high-temperate short-time (HTST) pasteurization at 72 °C for 15 and 30 s, and HPP applied at ~20 °C (**A**), ~30 °C (**B**), and ~40 °C(**C**). Error bars represent the standard deviations of duplicates.

Figure 7 represents the relative proportions (%) of α-Lac and β-Lg derived from Figure 6. The highest ratio of α-Lac to β-Lg (77:23) was achieved from the HPP treatment of 600 MPa at 51.7 °C for 5 min, whereas it was only 24:76 and 23:77 in HTST for 15 and 30 s, respectively (Figure 7). From the results obtained in this work, it is evident that the synergistic effect of HPP at elevated temperature induces a higher ratio of α-Lac to β-Lg, compared to the untreated (22:78) and HTST-treated α-Lac-added reconstituted IMF. The higher baroresistance of α-Lac, compared with β-Lg, is consistent with previous observations in milk [31,32,43]. This difference is considered to be due to the higher number of intramolecular disulfide bonds (4 in α-Lac and 2 in β-Lg) and to the presence of a free sulphydryl group in β-Lg [32,40]. Upon unfolding of β-Lg due to HPP, this free sulphydryl group interacts with proteins containing disulphide bonds (e.g., αs2-casein, k-casein, α-Lac, and β-Lg) through sulphydryl–disulfide interchange reactions resulting in aggregation. Moreover, unfolded α-Lac and β-Lg, which did not interact with other proteins, refold to their native forms on the release of pressure [31]. Therefore, the mechanistic approach of using HPP followed in this work explains the mechanism to achieve a final product with a massive reduction in the β-Lg portion, which subsequently would result in lowering the allergenicity and protein content.

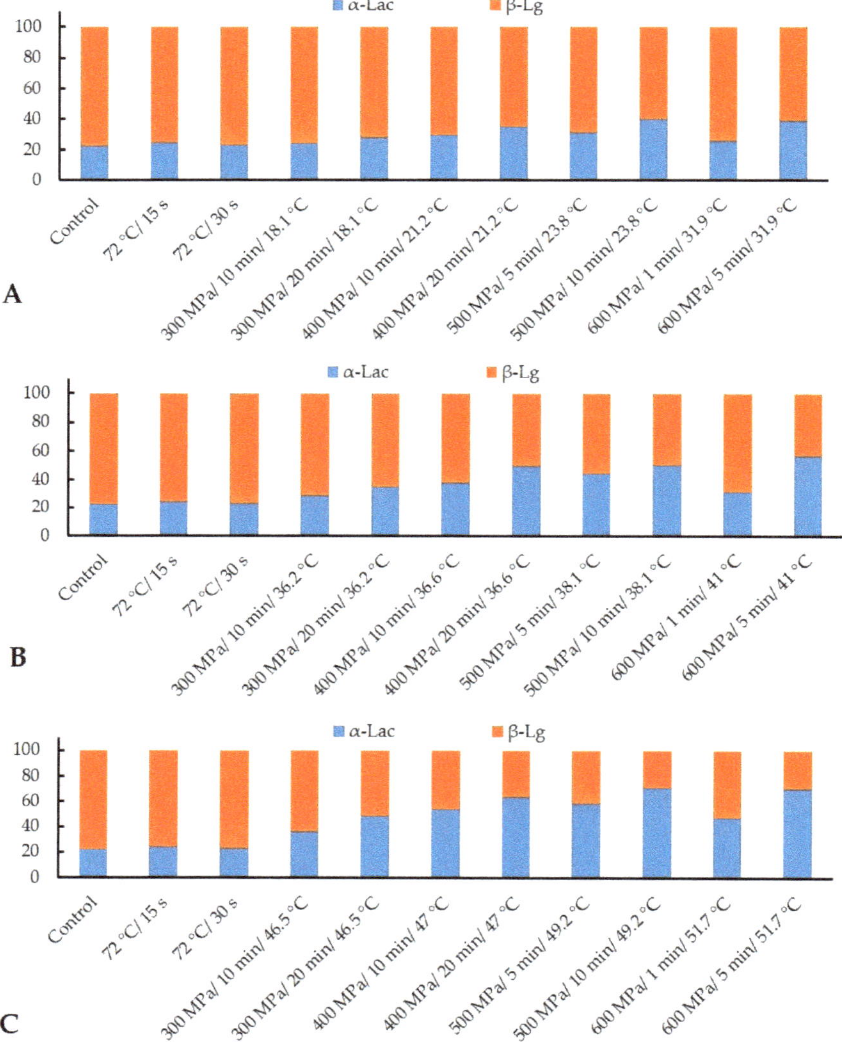

Figure 7. Relative proportions of α-Lac and β-Lg (%) after HTST at 72 °C for 15 and 30 s, and HPP applied at ~20 °C (**A**), ~30 °C (**B**), and ~40 °C (**C**). Total refers to the sum of α-Lac and β-Lg content.

The results found in this work show the potential route to develop an HPP-treated pasteurized RTF hypoallergenic formula because of having a higher ratio of α-Lac to β-Lg. In addition to this, such a formula would ensure the required amino acid balance in the treated product due to the α-Lac supplementation, which may also compensate the lower contribution of heavily denatured β-Lg in the amino acid profile. Thus, this work streamlines the possibility of manufacturing a hypoallergenic and low-protein pasteurized RTF formula. However, further investigation in post-treatment analysis (e.g., bioavailability, digestibility, amino acid profile, etc.) of HPP-treated formula is highly recommended to commercialize this research. Besides, the shorter shelf life at refrigerated conditions of this pasteurized product than that of the sterilized RTF formula would result in slower progress in gaining market. Moreover, HPP is still limited by its batch operation although the recent patent-pending concept of Hiperbaric, the HPP equipment manufacturer, to process liquid foods up to 10,000 L/h before bottling (aseptic packaging) is being considered as a promising innovation to address HPP's batch operation [44].

4. Conclusions

Our results demonstrated the synergistic effect of HPP and heat on the substantial reduction of β-Lg in reconstituted IMF added with α-Lac. Compared to HTST, the HPP treatment at 600 MPa for 5 min applied at 40.4 °C achieved the higher ratio of α-Lac to β-Lg. Overall, the pronounced reduction of β-Lg due to the combined effect of HPP and heat confirms the possibility to manufacture a hypoallergenic and low protein RTF formula, a niche product. Further investigation is necessary to explore the post-treatment effects on physicochemical and rheological properties, along with microbial studies.

Author Contributions: M.A.W. conceived and performed the experiments; analyzed the data and drafted the manuscript. M.F. reviewed, edited and supervised the overall study.

Funding: This work was supported by the Food Industry Enabling Technologies (FIET) programme funded by the New Zealand Ministry of Business, Innovation, and Employment (contract MAUX1402).

Acknowledgments: The assistance of Raymond Hoffman with HPP treatment is greatly appreciated. We acknowledge Davisco Foods International Inc. (Le Sueur, MN, USA) for providing the α-Lac powder.

Conflicts of Interest: The authors declare no conflict of interest.

References

1. World Health Organization. *Global Strategy for Infant and Young Child Feeding*; World Health Organization: Geneva, Switzerland, 2003; Available online: http://apps.who.int/gb/archive/pdf_files/WHA55/ea5515.pdf (accessed on 15 January 2019).
2. Eidelman, A.I.; Schanler, R.J.; Johnston, M.; Landers, S.; Noble, L.; Szucs, K.; Viehmann, L. Breastfeeding and the use of human milk. *Pediatrics* **2012**, *129*, e827–e841.
3. Martin, C.R.; Ling, P.R.; Blackburn, G.L. Review of infant feeding: Key features of breast milk and infant formula. *Nutrients* **2016**, *8*, 279. [CrossRef] [PubMed]
4. Adelman, S.W.; Schmeiser, K.N. Infant Formula Trade and Food Safety. *Appl. Econ. Financ.* **2018**, *6*, 1–10. [CrossRef]
5. Thompkinson, D.K.; Kharb, S. Aspects of infant food formulation. *Compr. Rev. Food Sci. Food Saf.* **2007**, *6*, 79–102. [CrossRef]
6. Andreas, N.J.; Kampmann, B.; Le-Doare, K.M. Human breast milk: A review on its composition and bioactivity. *Early Hum. Dev.* **2015**, *91*, 629–635. [CrossRef]
7. Fenelon, M.A.; Hickey, R.M.; Buggy, A.; McCarthy, N.; Murphy, E.G. *Whey Proteins in Infant Formula*. *Whey Proteins*; Academic Press: Cambridge, MA, USA, 2019; pp. 439–494.
8. Liu, Z.; Roy, N.C.; Guo, Y.; Jia, H.; Ryan, L.; Samuelsson, L.; Thomas, A.; Plowman, J.; Clerens, S.; Day, L.; et al. Human Breast Milk and Infant Formulas Differentially Modify the Intestinal Microbiota in Human Infants and Host Physiology in Rats–3. *J. Nutr.* **2015**, *146*, 191–199. [CrossRef] [PubMed]
9. Wojcik, K.Y.; Rechtman, D.J.; Lee, M.L.; Montoya, A.; Medo, E.T. Macronutrient analysis of a nationwide sample of donor breast milk. *J. Am. Diet. Assoc.* **2009**, *109*, 137–140. [CrossRef]
10. Von Berg, A.; Koletzko, S.; Grübl, A.; Filipiak-Pittroff, B.; Wichmann, H.E.; Bauer, C.P.; Reinhardt, D.; Berdel, D. The effect of hydrolyzed cow's milk formula for allergy prevention in the first year of life: The German Infant Nutritional Intervention Study, a randomized double-blind trial. *J. Allergy Clin. Immunol.* **2003**, *111*, 533–540. [CrossRef]

11. Nasirpour, A.; Scher, J.; Desobry, S. Baby foods: Formulations and interactions (a review). *Crit. Rev. Food Sci. Nutr.* **2006**, *46*, 665–681. [CrossRef]

12. Golkar, A.; Milani, J.M.; Vasiljevic, T. Altering allergenicity of cow's milk by food processing for applications in infant formula. *Crit. Rev. Food Sci. Nutr.* **2019**, *59*, 159–172. [CrossRef]

13. Lönnerdal, B. Infant formula and infant nutrition: Bioactive proteins of human milk and implications for composition of infant formulas. *Am. J. Clin. Nutr.* **2014**, *99*, 712S–717S. [CrossRef] [PubMed]

14. Sandström, O.; Lönnerdal, B.; Graverholt, G.; Hernell, O. Effects of α-lactalbumin–enriched formula containing different concentrations of glycomacropeptide on infant nutrition. *Am. J. Clin. Nutr.* **2008**, *87*, 921–928. [CrossRef] [PubMed]

15. Dewey, K.G.; Beaton, G.; Fjeld, C.; Lonnerdal, B.; Reeds, P.; Brown, K.H.; Heinig, M.J.; Ziegler, E.; Räihä, N.C.R.; Axelsson, I. Protein requirements of infants and children. *Eur. J. lin. Nutr.* **1996**, *50*, S119–S150.

16. Davis, A.M.; Harris, B.J.; Lien, E.L.; Pramuk, K.; Trabulsi, J. α-Lactalbumin-rich infant formula fed to healthy term infants in a multicenter study: Plasma essential amino acids and gastrointestinal tolerance. *Eur. J. Clin. Nutr.* **2008**, *62*, 1294–1301. [CrossRef] [PubMed]

17. Marsh, K.; Möller, J.; Basarir, H.; Orfanos, P.; Detzel, P. The Economic Impact of Lower Protein Infant Formula for the Children of Overweight and Obese Mothers. *Nutrients* **2016**, *8*, 18. [CrossRef] [PubMed]

18. Karlsland, A.P.; Axelsson, I.E.; Räihä, N.C. Protein and amino acid metabolism in three-to twelve-month-old infants fed human milk or formulas with varying protein concentrations. *J. Pediatr. Gastroenterol. Nutr.* **1998**, *26*, 297–304.

19. Schmidt, I.M.; Damgaard, I.N.; Boisen, K.A.; Mau, C.; Chellakooty, M.; Olgaard, K.; Main, K.M. Increased kidney growth in formula-fed versus breast-fed healthy infants. *Pediatr. Nephrol.* **2004**, *19*, 1137–1144. [CrossRef]

20. Kuhlman, C.F.; Lien, E.; Weaber, J.; O'callaghan, D. Infant Formula Compositions Comprising Increased Amounts of Alpha-Lactalbumin. U.S. Patent No. 6,913,778, 5 July 2005.

21. Lien, E.L. Infant formulas with increased concentrations of α-lactalbumin. *Am. J. Clin. Nutr.* **2003**, *77*, 1555S–1558S. [CrossRef]

22. Corredig, M.; Dalgleish, D.G. The mechanisms of the heat-induced interaction of whey proteins with casein micelles in milk. *Int. Dairy J.* **1999**, *9*, 233–236. [CrossRef]

23. Goyal, A.; Sharma, V.; Upadhyay, N.; Sihag, M.; Kaushik, R. High pressure processing and its impact on milk proteins: A review. *Res. Rev. J. Dairy Sci. Technol.* **2018**, *2*, 12–20.

24. Huppertz, T.; Fox, P.F.; Kelly, A.L. Dissociation of caseins in high pressure-treated bovine milk. *Int. Dairy J.* **2004**, *14*, 675–680. [CrossRef]

25. Needs, E.C.; Stenning, R.A.; Gill, A.L.; Ferragut, V.; Rich, G.T. High-pressure treatment of milk: Effects on casein micelle structure and on enzymic coagulation. *J. Dairy Res.* **2000**, *67*, 31–42. [CrossRef] [PubMed]

26. McCarthy, N.A.; Gee, V.L.; Hickey, D.K.; Kelly, A.L.; O'Mahony, J.A.; Fenelon, M.A. Effect of protein content on the physical stability and microstructure of a model infant formula. *Int. Dairy J.* **2013**, *29*, 53–59. [CrossRef]

27. Crowley, S.V.; Dowling, A.P.; Caldeo, V.; Kelly, A.L.; O'Mahony, J.A. Impact of α-lactalbumin: β-lactoglobulin ratio on the heat stability of model infant milk formula protein systems. *Food Chem.* **2016**, *194*, 184–190. [CrossRef] [PubMed]

28. Dion, S.; Duivestein, J.A.; Pierre, A.S.; Harris, S.R. Use of thickened liquids to manage feeding difficulties in infants: A pilot survey of practice patterns in Canadian pediatric centers. *Dysphagia* **2015**, *30*, 457–472. [CrossRef] [PubMed]

29. Tobin, J.T.; Fitzsimons, S.M.; Kelly, A.L.; Fenelon, M.A. The effect of native and modified konjac on the physical attributes of pasteurized and UHT-treated skim milk. *Int. Dairy J.* **2011**, *21*, 790–797. [CrossRef]

30. Kamarei, A.R. Refrigeration-Shelf-Stable Ultra-Pasteurized or Pasteurized Infant Formula. U.S. Patent No. 6,039,985, 20 March 2000.

31. Huppertz, T.; Fox, P.F.; Kelly, A.L. High pressure-induced denaturation of α-lactalbumin and β-lactoglobulin in bovine milk and whey: A possible mechanism. *J. Dairy Res.* **2004**, *71*, 489–495. [CrossRef]

32. Mazri, C.; Sánchez, L.; Ramos, S.J.; Calvo, M.; Pérez, M.D. Effect of high-pressure treatment on denaturation of bovine β-lactoglobulin and α-lactalbumin. *Eur. Food Res. Technol.* **2012**, *234*, 813–819. [CrossRef]

33. Sousa, S.G.; Delgadillo, I.; Saraiva, J.A. Human milk composition and preservation: Evaluation of high-pressure processing as a nonthermal pasteurization technology. *Crit. Rev. Sci. Nutr.* **2016**, *56*, 1043–1060. [CrossRef]

34. Wesolowska, A.; Sinkiewicz-Darol, E.; Barbarska, O.; Bernatowicz-Lojko, U.; Borszewska-Kornacka, M.K.; van Goudoever, J.B. Innovative Techniques of Processing Human Milk to Preserve Key Components. *Nutrients* **2019**, *11*, 1169. [CrossRef]

35. Ghani, A.A.; Farid, M.M. Numerical simulation of solid–liquid food mixture in a high pressure processing unit using computational fluid dynamics. *J. Food Eng.* **2007**, *80*, 1031–1042. [CrossRef]

36. Silva, F.V. High pressure processing of milk: Modeling the inactivation of psychrotrophic *Bacillus cereus* spores at 38–70 °C. *J. Food Eng.* **2015**, *165*, 141–148.

37. Farid, M.; Alkhafaji, S. Determination of an effective treatment temperature of chemical and biological reactions. *Food Bioprocess Technol.* **2012**, *5*, 147–154. [CrossRef]

38. Anema, S.G.; McKenna, A.B. Reaction kinetics of thermal denaturation of whey proteins in heated reconstituted whole milk. *J. Agric. Food Chem.* **1996**, *44*, 422–428. [CrossRef]

39. Anema, S.G.; Stockmann, R.; Lowe, E.K. Denaturation of β-lactoglobulin in pressure-treated skim milk. *J. Agric. Food Chem.* **2005**, *53*, 7783–7791. [CrossRef] [PubMed]

40. Hinrichs, J.; Rademacher, B.; Kessler, H.G. Reaction kinetics of pressure-induced denaturation of whey proteins. *Milchwissenschaft* **1996**, *51*, 504–509.

41. Bogahawaththa, D.; Chandrapala, J.; Vasiljevic, T. Thermal denaturation of bovine β-lactoglobulin in different protein mixtures in relation to antigenicity. *Int. Dairy J.* **2019**, *91*, 89–97. [CrossRef]

42. Marciniak, A.; Suwal, S.; Britten, M.; Pouliot, Y.; Doyen, A. The use of high hydrostatic pressure to modulate milk protein interactions for the production of an alpha-lactalbumin enriched-fraction. *Green Chem.* **2018**, *20*, 515–524. [CrossRef]

43. Lopez-Fandino, R.; Carrascosa, A.V.; Olano, A. The effects of high pressure on whey protein denaturation and cheese-making properties of raw milk. *J. Dairy Sci.* **1996**, *79*, 929–936. [CrossRef]

44. Hiperbaric 1050 Bulk. Available online: https://www.hiperbaric.com/en/hiperbaric1050bulk (accessed on 30 September 2019).

 *foods*

Article

Yoghurt-Type Gels from Skim Sheep Milk Base Enriched with Whey Protein Concentrate Hydrolysates and Processed by Heating or High Hydrostatic Pressure

Lambros Sakkas [1], Maria Tzevdou [2], Evangelia Zoidou [1], Evangelia Gkotzia [1], Anastasis Karvounis [1], Antonia Samara [1], Petros Taoukis [2] and Golfo Moatsou [1,*]

[1] Laboratory of Dairy Research, Department of Food Science and Human Nutrition, Agricultural University of Athens, Iera Odos 75, 118 55 Athens, Greece
[2] Laboratory of Food Chemistry and Technology, School of Chemical Engineering, National Technical University of Athens (NTUA), Laboratory of Food Chemistry and Technology, 5 Heroon Polytechniou Str., Zografos, 15780 Athens, Greece
* Correspondence: mg@aua.gr; Tel.: +30-2105294630

Received: 30 June 2019; Accepted: 9 August 2019; Published: 12 August 2019

Abstract: An objective of the present study was the enrichment of skim sheep yoghurt milk base with hydrolysates (WPHs) of whey protein concentrate (WP80) derived from Feta cheesemaking. Moreover, the use of high hydrostatic pressure (HP) treatment at 600 MPa/55 °C/10 min as an alternative for heat treatment of milk bases, was studied. In brief, lyophilized trypsin and protamex hydrolysates of WP80 produced under laboratory conditions were added in skim sheep milk. The composition and heat treatment conditions were set after the assessment of the heat stability of various mixtures; trisodium citrate was used as a chelating agent, when needed. According to the results, the conditions of heat treatment were more important for the physical properties of the gel than the type of enrichment. High pressure treatment resulted in inferior gel properties, irrespective of the type of enrichment. Supplementation of skim sheep milk with whey protein hydrolysates at >0.5% had a detrimental effect on gel properties. Finally, skim sheep milk base inoculated with fresh traditional yoghurt, resulted in yoghurt-type gels with high counts of *Lb. delbrueckii* subsp. *bulgaricus* and *Str. thermophilus* -close to the ideal 1:1- and with a high ACE inhibitory activity >65% that were not essentially affected by the experimental factors.

Keywords: high hydrostatic pressure; whey protein hydrolysates; sheep milk; yoghurt; ACE inhibitory activity; gel properties; heat stability; traditional yoghurt starter; biofunctionality

1. Introduction

The manufacture of low-fat, high-protein fermented milks is one of the current trends in food technology. High-protein, high-mineral set-style yoghurt manufactured from partially defatted or skim sheep milk can be stable throughout four weeks of storage. However, the fat removal reduces the water holding capacity and the firmness of the skim yoghurt in comparison to the reduced-fat counterpart [1]. An objective of the present study was the enrichment of the skim sheep milk base with hydrolysates (WPHs) of whey protein concentrate (WP80) derived from Feta cheesemaking. The hypothesis was that this intervention could modify the texture and increase the biofunctionality, in terms of angiotensin converting enzyme inhibitory activity (ACE-IA), in accordance to our recent findings for reduced-fat cow milk yoghurt [2]. Fortification or enrichment of yoghurt cow milk base with WPC has been widely studied [3,4], but similar interventions in small ruminants yoghurt milk are scarce and have been applied in goat milk [5–7].

Hydrolysis of whey or individual whey proteins or whey protein concentrates by various enzymes results in mixtures of proteins and peptides known as whey protein hydrolysates (WPHs). WPHs can exhibit improved biofunctionality and modified physical properties compared to the substrate [4,8,9]. The use of whey protein hydrolysates (WPHs) in the cow milk yoghurt base has been studied in regard to the growth of probiotics [10,11] or biofunctional and textural properties [2,12].

An important and essential step of the manufacture of fermented milks is the heat treatment of the milk base. The heat stability of sheep milk is lower than cow milk due to differences in the casein micelle structure and mineral content [13,14]. Therefore, another objective of the present study was the use of high hydrostatic pressure (HP) as an alternative for the heat treatment of yoghurt milk bases. HP processing has been associated with variable effects on yoghurt properties that depend on the composition of cow milk base [15–17]. However, information for the effect of HP-treatment on sheep milk that could affect yoghurt manufacture and properties is limited [18–23].

2. Materials and Methods

2.1. Hydrolysis of Whey Protein Concentrate

The substrate was a 5% (*w*/*w*) aqueous dilution of whey protein concentrate powder with approximately 80% (*w*/*w*) protein content (WP80) derived from sheep/goat Feta cheese whey (Epirus Protein SA, Ioannina, Greece). The composition of WP80 (Table 1) was determined as described by Zoidou et al. [7]. The aqueous dilution of WP80 remained overnight at 4 °C–6 °C for proper hydration of the powder. Hydrolysis was carried out by means of trypsin and Protamex. The ratio enzyme to protein was 0.25%, and 0.1% for trypsin and protamex, respectively. Incubations were carried out at 50 °C without pH adjustment at pH 6.2–6.3, which was the pH of the WP80 dilution. After 60 min, the pH was 5.6–6.1 and hydrolysis was stopped by heat treatment at 68 °C for 10 min, which are the conditions for batch pasteurization of milk. Hydrolysates were powdered by lyophilization. Trypsin and protamex hydrolysates were symbolized as WPHt and WPHp. Hydrolysis conditions resulted from previous experiments (data not shown).

Table 1. Composition of whey protein concentrate (WP80) g/100 g used as substrate for the production of whey protein hydrolysates (WPHs).

Component	Ca	Mg	K	Na	P	CMP	α-la	SA	β-lg	TNP
g/100 g	0.323	0.065	0.259	0.205	0.214	12.75	14.02	2.63	46.55	75.95

CMP, caseinomacropeptide; α-la, α-lactalbumin, SA, serum albumin; β-lg, β-lactoglobulin; TNP, total native protein.

2.2. Assessment of Heat Stability of Various Sheep Milk Bases

The aim of this experimental section was the selection of heat stable milk bases that could be used for the manufacture of yoghurt-type gels. Raw sheep milk was defatted by centrifugation at 3000× *g* for 20 min at 40 °C. Skim sheep milk (0.1% fat) was mixed with variable quantities (0.5%–2% *w*/*w*) of WP80, WPHt and WPHp powders with or without trisodium citrate at 0.2% (*w*/*w*), i.e., approximately 7 mM. Experiments were performed in duplicate and three replicates of each mixture were prepared in each experiment. Mixtures were kept overnight at 5 °C–6 °C for hydration. Then, two of the replicates were heated at 70 °C, 75 °C, 80 °C, 85 °C, 90 °C and 95 °C for 5 min. At the end of the heat treatment, the milk mixtures were cooled down immediately at room temperature using an ice-bath. One of the triplicates was not heated (control). Heat-treated (HT) and control mixtures were centrifuged at 10,000× *g* for 10 min. Appearance of sediment indicated heat instability.

2.3. Heat- and High Pressure- Treatment of Selected Skim Sheep Milk Bases

Raw ovine milk was defatted by means of a lab-scale milk fat separator. The composition and treatments of milk bases used in the subsequent yoghurt-type gel experiments are shown in Table 2.

The criterion for their selection was the heat stability (Section 2.2). Heat treatments (HT) were carried out under batch conditions. The same mixtures without trisodium citrate were processed by high pressure (HP) as follows. Treatments were performed using a laboratory-scale HP system with a maximum operating pressure of 1000 MPa (Food Pressure Unit FPU 1.01, Resato International BV, Roden, Netherlands), consisting of an HP unit with a pressure intensifier, an HP vessel of 1.5 L and a multi-vessel system consisting of six vessels of 42 mL capacity each. All HP vessels are surrounded by a water circulating jacket connected to a temperature control system. The pressure-transmitting fluid used was the polyglycol ISO viscosity class VC 15 (Resato International BV, Netherlands). Milk bases were put into a multilayer (PP, foil, PE) packaging and placed in the 1.5 L chamber for processing. The desired value of pressure was set and, after pressure build-up (ca. 20 MPa·s^{-1}), the pressure vessel was isolated; this point defined the zero time of the process. The pressure of the vessel was released after a preset time interval (10 min pressurization time) by opening the pressure valve (release time <3 s). The initial temperature increase during the pressure build-up (ca. 3 °C per 100 MPa) was taken into consideration in order to achieve the desired operating temperature. The pressure and temperature were constantly monitored (intervals of 1 s) and recorded during the process. All samples were pressurized at 619 ± 7 MPa and 55.2 ± 1.3 °C for a process time of 10 min. After processing, samples were kept overnight at 4 °C. The heat and high-pressure experiments were performed in duplicate.

Table 2. Composition and treatments of skim sheep milk bases. Means of two experiments ± standard deviation.

Milk Base	Powder % (*w/w*)	Trisodium Citrate	Treatment	pH before Treatment	pH after Treatment
ϒ0 [1]-HT	-	-	95 °C/5 min	6.71 ± 0.09	6.36 ± 0.07
ϒ0 [1]-HP	-	-	600 MPa/55 °C/10 min	6.66 ± 0.01	6.54 ± 0.16
ϒWP80-HT	WP80 [2]/1%	0.2%	90 °C/5 min	6.78 ± 0	6.52 ± 0.05
ϒWP80-HP		-	600 MPa/55 °C/10 min	6.63 ± 0.02	6.42 ± 0.10
ϒWPHp0.5-HT	WPHp [3]/0.5%	0.2%	85 °C/5 min	6.85 ± 0.05	6.59 ± 0.03
ϒWPHp0.5-HP		-	600 MPa/55 °C/10 min	6.61 ± 0	6.41 ± 0.07
ϒWPHt0.5-HT	WPHt [4]/0.5%	0.2%	85 °C/5 min	6.84 ± 0.05	6.55 ± 0.03
ϒWPHt0.5-HP		-	600 MPa/55 °C/10 min	6.60 ± 0.04	6.43 ± 0.11
ϒWPHt1-HT	WPHt [4]/1%	0.2%	75 °C/5 min	6.80 ± 0.03	6.46 ± 0.09
ϒWPHt1-HP		-	600 MPa/55 °C/10 min	6.61 ± 0.04	6.39 ± 0.10

[1] defatted sheep milk (control); [2] whey protein concentrate powder with approximately 80% (w/w) protein content (WP80) derived from sheep/goat Feta cheese whey; [3] trypsin hydrolysate of WP80; [4] protamex hydrolysate of WP80.

2.4. Manufacture of Yoghurt-type Gels

After treatments, the temperature of the mixtures of Table 2 was adjusted to 43 °C, which was the inoculation temperature. Fresh Greek traditional sheep yoghurt with the characteristic top fat layer [24] was used as a starter, at a ratio of 1.5% (*v/v*). Portions of 100 mL of inoculated milk bases were poured in sterilized containers, in which they were incubated at 42 °C, until pH 4.7. Incubation was stopped by immediate cooling. At the end of cooling, the yoghurt-type gels had pH 4.6. The average duration of incubation was 2 h and 30 min and 2 h and 18 min for HT and HP treated mixtures, respectively.

2.5. Analyses

All analyses were performed as previously described [1,2,7]. In brief, gross composition, yoghurt bacteria counts, water holding capacity (WHC), firmness/cohesiveness and ACE-IA determinations were carried out by means of the Milkoscan, colony count technique, centrifugation, texture profile analysis and RP-HPLC, respectively.

2.6. Statistical Analysis

The analysis of variance (ANOVA) was used to analyze the effect of the mixture composition (milk base) and of the treatment on the characteristics of the yoghurt-type gels and for the differences among the means of the Least Significant Difference test was used (LSD, $p < 0.05$). Statistical analysis was carried out by means of the Statgraphics, Centurion V (Manugistics Inc., Rockville, MA 20852, USA).

3. Results and Discussion

Sodini et al. [3] summarized that the rheological properties and microstructure of yoghurt are related to the heating conditions of the milk base from 75 °C for 1 min–5 min to 95 °C for 5 min–10 min; the later conditions result in >99% denaturation of the β-lactoglobulin (β-lg). The heat treatment induces the formation of various types of complexes, which are key factors for the configuration of acid milk gels and yoghurt: i. between casein micelles and denatured whey proteins, ii. between κ-casein and β-lg and iii. between denatured whey proteins [25]. Their extent and distribution are differentiated within the range from pH 6.5 to pH 6.7. At low pH, the denatured whey proteins bind onto the casein micelles thus increasing the particle size while at higher pH they participate in soluble complexes with solubilized κ-casein [26].

From Table 2, it is evident that only the control skim sheep milk base was stable under the usual heating conditions for the yoghurt manufacture –i.e., 95 °C/5 min. The addition of WPC or WPHs even at low level decreased the heat stability and trisodium citrate addition was necessary. The heat stability of sheep milk is lower than that of cow milk because ovine micelles are more mineralized, contain more β-casein, are less hydrated than their bovine counterparts and their size increase substantially during heating. In fact, heating at >80 °C causes an increase of the ovine micelle size by >50%, which along with the high casein content favour micelle-micelle interaction and aggregation [13,14]. The later phenomenon is expected to be more pronounced in the skim sheep milk of the present experiments due to the increase of protein concentration caused by the removal of fat. Moreover, the calcium content of WPC and WPHs may also adversely affect the heat stability of the mixtures of Table 2. Excessive calcium before heat treatment results in the formation of large aggregates that decrease the heat-stability while calcium combined with acidification induce the gelation of denatured whey protein polymers [27].

Therefore, both the low pH and the elevated calcium content of the milk bases before acidification could be responsible for the poor heat stability of the sheep milk bases in the experiments of Section 2.2. Milder heating conditions and the addition of trisodium citrate before heat treatment were used for the enriched skim milk bases. As shown in Table 2, the pH of milk bases supplemented with approximately 7 mM sodium citrate was higher compared to the control before and after heat treatment, on average by 0.12 pH and 0.16 pH units, respectively. The addition of sodium citrate in sheep or goats milk enhances their heat stability by linking with ionic calcium and solubilizing both the colloidal calcium phosphate and the calcium linked to the phosphoseryl residues. Moreover, sodium citrate increases the milk pH by approximately 0.1 units resulting in the increase of small-sized casein micelles and of their negative charge that do not favour heat aggregation [13,14]. In regard to cow milk yoghurt gels, using >25 mM trisodium citrate causes detrimental disruption of casein micelles, while the addition of 10 mM–20 mM trisodium citrate improves the texture by chelating calcium. Calcium chelation induces solublization of colloidal calcium phosphate, thus enhancing the formation of crosslinks [28]. The milk bases were stable under the conditions of HP treatment, which reduced the pH by 0.18 units, on average.

3.1. Properties of Yoghurt-type Gels

The effect of the experimental factors on the physical and compositional properties of yoghurt-type gels was analyzed by the multifactor ANOVA and it is shown in Table 3.

Table 3. Effect of experimental factors on the properties of yoghurt-type gels, expressed as *p*-values.

Properties	A: composition	B: treatment	C: days	A × B	A × C	B × C
pH	0.119	0.634	0	0.034	0.288	0.023
WHC (%)	0	0	0.07	0	0.928	0.371
FIRMNESS (N)	0	0	0.007	0	0.753	0.157
COHESIVENESS	0	0	0.999	0	0.985	0.799
TOTAL SOLIDS (%)	0.001	0.001	-	0.170	-	-
PROTEIN (%)	0.003	0.144	-	0.057	-	-

It is evident (Table 3) that the physical properties (water holding capacity, firmness and cohesiveness) were affected significantly ($p < 0.05$) by the type of enrichment (composition) of the milk base and by the processing (heat- or high pressure-treatment). Moreover, there was a significant ($p < 0.05$) combined effect of these factors (A × B) on the physical properties. The days of storage affected significantly ($p < 0.05$) the pH and the firmness of gels.

The physical and compositional properties of the yoghurt-type gels treated by heat or by high-pressure are presented in Table 4. The factor "composition"- that is the composition of the milk base mixture—was related to statistically significant differences ($p < 0.05$) mostly in the HT group. These differences can be assigned to: i. different heat treatments, and ii. differences in composition. It is evident that at day three, the YWPHt1 base treated at 75 °C for 5 min, i.e., under the mildest conditions, was significantly differentiated among the HT group in regard to the water holding capacity (WHC), firmness and cohesiveness. The extend of whey protein denaturation in YWPHt1 was expected to be the lowest, which is consistent with its lowest WHC. Extended heat denaturation of β-lg enhances the capability of the casein network to immobilize the serum [3]. On the other hand, in high-protein yoghurts the reduction of heating temperature from 95 °C to 75 °C for 5 min reduces firmness but improves the sensory properties [29]. Despite the highest total solids and protein content of YWPHt1, its firmness was the lowest in accordance to the above-mentioned effect of heating on the formation and distribution of complexes. It has to be noticed that the enrichment of skim milk - i.e., higher total solids and protein contents - and heat treatment at 85 °C or 90 °C for 5 min resulted in significantly higher ($p < 0.05$) WHC and firmness for YWP80, YWOHp0.5 and WPHt0.5 compared to the control Y0. Therefore, the supplementation with WP80 at 1% and with WPHs at 0.5% counteracted the effect of milder heating conditions. The addition of whey proteins in the yoghurt milk base increases also the ratio whey protein to casein that in turn increases WHC and affect also the viscoelasticity and flow behavior; the latter is related to heating conditions [3]. In particular, the whey protein to the casein ratio has been demonstrated as a crucial factor for the structure of non-fat stirred yoghurts [30]. The enrichment with WPHs resulted in lower WHC and firmness compared to the intact WP80. Apparently, due to hydrolysis, the WPHt and WPHp contained less intact native whey proteins, which are key components for the crosslinking within the yoghurt gel matrix, as reported above. Moreover, the solubility of a WPH may be reduced due to the exposure of hydrophobic areas of the molecules [9].

Table 4. Properties of yoghurt-type gels made from skim sheep milk bases enriched with WPHs.

Properties	Treatment	Y0	YWP80	YWPHp0.5	YWPHt0.5	YWPHt1
Day 3						
pH	HT [1]	4.35 ± 0.05 a	4.44 ± 0.02 b	4.33 ± 0.03 a	4.32 ± 0.06 a	4.38 ± 0.04 a, b
	HP [2]	4.41 ± 0.06	4.35 ± 0.08	4.29 ± 0.06	4.32 ± 0.09	4.32 ± 0.14
WHC (%)	HT	33.21 ± 1.92 b	45.31 ± 1.53 d, B	40,91 ± 2.13 c, B	37.23 ± 1.38 c, B	27.53 ± 2.84 a
	HP	28.52 ± 2.60	29.09 ± 2.21 A	26.72 ± 2.29 A	28.65 ± 1.43 A	28.69 ± 3.52
FIRMNESS	HT	1.40 ± 0.22 b, B	1.78 ± 0.21 b, c, B	1.87 ± 0.24 c, B	1.59 ± 0.14 b, c, B	0.66 ± 0.29 a
(N)	HP	0.70 ± 0.03 c, A	0.35 ± 0.06 a, A	0.52 ± 0.07 b, A	0.69 ± 0.04 c, A	0.57 ± 0.09 b, c
COHESIVENESS	HT	0.426 ± 0.019 a, A	0.421 ± 0.026 a, A	0.422 ± 0.014 a, A	0.449 ± 0.023 a	0.541 ± 0.039 b
	HP	0.540 ± 0.050 a, b, B	0.612 ± 0.019 b, B	0.506 ± 0.029 a, B	0.506 ± 0.021 a	0.546 ± 0.052 a, b
Day 10						
pH	HT	4.15 ± 0.07 a *	4.23 ± 0.05 a, b *	4.18 ± 0.03 a *	4.19 ± 0.04 a, b *	4.26 ± 0.02 b *
	HP	4.30 ± 0.04	4.23 ± 0.05	4.22 ± 0.04	4.23 ± 0.02	4.21 ± 0.03
WHC (%)	HT	36.61 ± 1.82 a, b, B	46.49 ± 3.52 c, B	39.74 ± 3.92 b	38.42 ± 3.58 b	30.85 ± 3.96 a
	HP	31.64 ± 1.28 A	29.87 ± 1.83 A	33.85 ± 3.00	31.90 ± 2.14	29.53 ± 2.35
FIRMNESS	HT	1.49 ± 0.36 b	2.20 ± 0.23 c, B	2.04 ± 0.33 c, B	1.83 ± 0.14 b, c, B	0.96 ± 0.15 a
(N)	HP	0.74 ± 0.03 c	0.38 ± 0.02 a, A	0.55 ± 0.03 b, A	0.77 ± 0.01 c, A	0.79 ± 0.02 c
COHESIVENESS	HT	0.429 ± 0.018 a, b	0.408 ± 0.028 a, A	0.438 ± 0.018 a, b	0.456 ± 0.021 b	0.507 ± 0.033 c
	HP	0.527 ± 0.057 a, b	0.646 ± 0.036 b, B	0.454 ± 0.043 a	0.502 ± 0.035 a, b	0.595 ± 0.99 a, b
TOTAL	HT	11.22 ± 0.28 a	12.05 ± 0.30 b	12.10 ± 0.25 b	12.15 ± 0.30 b	12.53 ± 0.21 c, B
SOLIDS (%)	HP	11.25 ± 0.21	11.98 ± 0.25	11.80 ± 0.35	11.58 ± 0.32	11.45 ± 0.21 A
PROTEIN	HT	4.73 ± 0.18 a	5.28 ± 0.23 b, c	5.15 ± 0.17 b	5.13 ± 0.24 b	5.59 ± 0.20 c
(%)	HP	4.80 ± 0.14 a	5.40 ± 0.14 b	4.93 ± 0.18 a	5.18 ± 0.18 a, b	4.98 ± 0.25 a, b

Means of two experiments ± standard deviation. Symbols as in Table 2; [1] heat treatment conditions indicated in Table 2; [2] high hydrostatic pressure treatment at 600 MPa/55 °C/10 min. a–d lowercase letters indicate statistically significant differences (LSD, $p < 0.05$) within each type of treatment, i.e., within rows; A–B, indicate significant differences ($p < 0.05$) between heat treatments (HT) and high pressure (HP) treatments; * indicates significant differences ($p < 0.05$) between three and 10 days.

Similar findings have been reported by other researchers. The use of the WPHt of the present study in reduced-fat cow milk base did not influence WHC but dramatically affected firmness, while the use of commercial WPHs of bovine origin had the opposite effect [2]. The addition of commercial WPHs in reduced-fat cow milk base at levels lower than 0.4% (*w/v*) affected negatively the texture of yoghurts due to less cross-linked microstructure [11]. The use of a tryptic WPH in buffalo milk at a ratio of 3% decreased the firmness and increased the syneresis of the sweetened yoghurt made there from [12].

The observation that the heating conditions were more important than the enrichment of the sheep milk base coincides with the findings for the HP-group of yoghurt-type gels. Less differences were observed between the HP milk bases (Table 2). These differences can be attributed solely to the enrichment since all were treated under the same conditions. Similar HP treatments of full-fat sheep milk induced >90% denaturation of β-lg and substantial reduction of α-la [22]. Firmness of the HP-group was significantly lower compared to the HT-group (Table 4) and this holds true for the control non-supplemented milk base Y0. WHC was not affected but firmness of YWP80 supplemented with 1% WP80 was significantly the lowest, half that of the control Y0. The opposite was true for cohesiveness. From the literature reviews [3,16,17] comes out that according to several studies, the HP treatment improve firmness and WHC. However, there are also opposite reports that coincide with our findings [15,29]. Similarly to the present study, the HP treatment of the cow milk base supplemented with whey protein concentrates and isolates resulted in weaker and less firm acid gels compared to heat treatment due to differences in the complexation of denatured β-lg [31,32]. The study of the effect of HP on reconstituted skim milk powder base combined with heating and in comparison to heat-treatment suggested that it can be used for the production of high protein drinking yoghurts of low viscosity [33]. Interestingly, the incorporation of WPHs in the HP treated milk bases had less detrimental effects on physical properties than the addition of non-hydrolyzed WP80. Again, a possible explanation is a favourable change of solubilization and the interactions between peptides and proteins, induced by partial hydrolysis [9].

The HP treatment of homogenized pasteurized full-fat sheep milk at 500 MPa/55 °C increase the firmness and the WHC of yoghurt compared to the typical 95 °C/5 min heat treatment [19]. The effect of HP on skim sheep milk has not been reported, but it is reasonable to expect that its high protein content could affect its behavior under various HP conditions. Additionally, the particularities of sheep milk and casein should be taken into consideration. Our previous study [23] showed that HP conditions similar to the present study affected the rennet clotting behavior of full-fat sheep milk in a different manner compared to cow milk. HP at 600 MPa decreased the size of ovine micelle by 40% [21]. At 400 MPa the κ-casein was extensively solubilized by >80%, much higher than the 22% observed in ovine milk [18]. Therefore, it could be assumed that soluble κ-casein/β-lg complexes were favoured under the applied HP conditions impairing thus crosslinking between micelles and β-lg and consequently the gel microstructure.

3.2. Biofunctional Characteristics

The thermophilic counts are "components" of yoghurts with biological value and along with particular substances such as proteins, peptides, vitamins, specific lipid compounds configure the biofunctionality of this type of fermented milks. Thermophilic bacilli and cocci log counts—estimated according to the ISO 7889-IDF 117 standard [34] - and the percentage of angiotensin converting enzyme inhibitory activity (ACE-IA)—often called anti-hypertensive potential- are shown in Table 5.

According to the results, both the enrichment and the treatment of the milk base did not affect the counts of *Lb. bulgaricus* and *Str. thermophilus*, which were high and very close to the suggested 1:1 ratio, after 10 days of storage. Of particular interest are the high numbers of bacilli which are considered very beneficial for the gastrointestinal system. The modern trend for yoghurts and fermented milks with mild characteristics has suppressed the presence and the viability of *Lactobacillus delbrueckii* subsp. *bulgaricus* in yoghurt starters to avoid the development of excessive acidity during storage. In our previous experiments carried out with commercial starters [1,7], we estimated less than four log cfu/g thermophilic bacilli whereas cocci counts were close to nine log cfu/g. Very low or even zero lactobacilli counts have also been estimated in market yoghurts [35]. Apparently, the biofunctional profile of the yoghurt-type gels of Table 5 is related to the Greek traditional sheep yoghurt used as a starter that contained 8.52 log cfu/g *Lb. bulgaricus* and 6.82 log cfu/g *Str. thermophilus*, while its ACE-IA was 57.5%.

ACE-IA in the present gels was high coinciding with the high bacilli counts and the high % ACE-IA of the starter. In our previous studies [1,7] the ACE-IA of the skim sheep milk set-style yoghurt ranged from 22% to 25%. In the present study, ACE-IA was not affected by the type of treatment and statistical differences were observed only between different mixtures of the HT-group. In general, the enrichment of dairy products with whey-derived peptides is expected to "add" biofunctional ACE-I peptides [36]. Nevertheless, in the present experiments the supplementation of the skim milk base decrease slightly this activity, indicating the importance of the viable counts for this category of dairy products.

Table 5. Biofunctional characteristics of yoghurt-type gels made from skim sheep milk bases enriched with WPHs after ten days of storage.

Parameter	Treatment	Y0	YWP80	YWPHp0.5	YWPHt0.5	YWPHt1	Factors/p-Values [5]		
							A: composition	B: treatment	A × B
Day 10									
Lb. bulgaricus [3]	HT [1]	8.20 ± 0.38	8.08 ± 0.25	8.26 ± 0.12 B	8.40 ± 0.12	8.23 ± 0.21	0.294	0	0.482
(log cfu/g)	HP [2]	7.39 ± 0.34	7.83 ± 0.21	7.64 ± 0.32 A	7.92 ± 0.28	7.91 ± 0.33			
Str. thermophilus [4]	HT	6.74 ± 0.41	7.28 ± 0.43	7.36 ± 0.33	7.52 ± 0.61	7.36 ± 0.81	0.327	0.965	0.564
(log cfu/g)	HP	6.82 ± 0.71	7.91 ± 0.77	7.38 ± 0.45	6.79 ± 0.86	7.43 ± 0.67			
ACE-IA (%)	HT	76.40 ± 1.95 b	73.97 ± 3.23 a, b	73.43 ± 2.04 a, b	70.9 ± 1.15 a B	73.53 ± 3.52 a, b	0.019	0.009	0.542
	HP	73.25 ± 4.60	74.2 ± 4.95	69.3 ± 2.97	64.65 ± 3.32 A	68.15 ± 2.90			

Means of two experiments ± standard deviation. Symbols as in Table 2; [1] heat treatment conditions indicated in Table 2; [2] high hydrostatic pressure treatment at 600 MPa/55 °C/10 min; [3] *Lactobacillus delbrueckii* subsp. *bulgaricus* [34]; [4] *Streptococcus thermophilus* [34]; [5] two-way ANOVA. a–b, lowercase letters indicate statistically significant differences (LSD, $p < 0.05$) within each type of treatment, i.e., within rows; A–B, indicate significant differences ($p < 0.05$) between HT and HP treatments.

4. Conclusions

The conditions of the heat treatment of the milk base were more important than the type of enrichment for the physical properties of the gel. These conditions were determined by the heat stability of the mixtures of "versatile" skim sheep milk with WPC or WPHs. The high pressure treatment at conditions that denature almost totally the β-lactoglobulin as an alternative to heating resulted in inferior gel properties, irrespective of the type of enrichment. Supplementation of skim sheep milk with whey protein hydrolysates at > 0.5% had a detrimental effect on gel properties. Finally, the sheep milk base along with the use of fresh traditional yoghurt as a starter, resulted in yoghurt-type gels with high counts of *Lb. delbrueckii* subsp. *bulgaricus* and *Str. thermophilus* -close to the ideal 1:1- and with high ACE-IA, which were not essentially affected by the experimental factors.

Author Contributions: Conceptualization, G.M.; Methodology, G.M., P.T. and M.T.; Software, Validation, G.M., E.Z. and L.S.; Formal Analysis, L.S., M.T., E.G., A.K., A.S. and E.Z.; Investigation, G.M. and P.T.; Resources, G.M. and P.T.; Data Curation, L.S., G.M., and E.Z.; Writing—Original Draft Preparation, G.M.; Writing—Review and Editing, G.M.; Visualization, L.S., G.M. and M.T..; Supervision, G.M. and P.T.; Project Administration, G.M. and P.T.; Funding Acquisition, M.G. and P.T.

Acknowledgments: This work was supported by the Greek General Secretariat for Research and Technology in the framework of the program SYNERGASIA II_2011 (code 530 11SYN_2_1265).

Conflicts of Interest: The authors declare no conflict of interest.

Abbreviations

ACE-IA	angiotensin-converting enzyme- inhibitory activity
CMP	caseinomaropeptide
cfu	colony forming unit
RP-HPLC	reversed-phase, high performance liquid chromatography
SA	serum albumin
TNP	total native protein
α-la	α-lactalbumin
β-lg	β-lactoglobulin

References

1. Moschopoulou, E.; Sakkas, L.; Zoidou, E.; Theodorou, G.; Sgouridou, E.; Kalathaki, C.; Liarakou, A.; Chatzigeorgiou, A.; Politis, I.; Moatsou, G. Effect of milk kind and storage on the biochemical, textural and biofunctional characteristics of set-type yoghurt. *Intern. Dairy J.* **2018**, *77*, 47–55. [CrossRef]
2. Roumanas, D.; Moatsou, G.; Zoidou, E.; Sakkas, L.; Moschopoulou, E. Effect of enrichment of bovine milk with whey proteins on biofunctional and rheological properties of low fat yoghurt-type products. *Curr. Res. Nutr. Food. Sci.* **2016**, *4*. [CrossRef]
3. Sodini, I.; Remeuf, F.; Haddad, S.; Corrieu, G. The relative effect of milk base, starter, and process on yogurt texture: A review. *Crit. Rev. Food Sci. Nutr.* **2004**, *44*, 113–137. [CrossRef] [PubMed]
4. Karam, M.C.; Gaiani, C.; Hosri, C.; Burgain, J.; Scher, J. Effect of dairy fortification on yogurt textural and sensorial properties: A review. *J. Dairy Res.* **2013**, *80*, 400–409. [CrossRef]
5. Bruzantin, F.P.; Daniel, J.L.P.; da Silva, P.P.M.; Spoto, M.H.R.F. Physicochemical and sensory characteristics of fat-free goat milk yogurt added to stabilizers and skim milk powder fortification. *J. Dairy Sci.* **2016**, *99*, 1–9. [CrossRef] [PubMed]
6. Gursel, A.; Gursoy, A.; Anli, E.A.K.; Budak, S.O.; Aydemir, S.; Durlu-Ozkaya, F. Role of milk protein-based products in some quality attributes of goat milk yogurt. *J. Dairy Sci.* **2016**, *99*, 2694–2703. [CrossRef] [PubMed]
7. Zoidou, E.; Theodorou, S.; Moschopoulou, E.; Sakkas, L.; Theodorou, G.; Chatzigeorgiou, A.; Politis, I.; Moatsou, G. Set-style yoghurts made from goat milk bases fortified with whey protein concentrates. *J. Dairy Res.* **2019**, in press.
8. Abd El-Salam, M.H.; El-Shibiny, S. Preparation, properties, and uses of enzymatic milk protein hydrolysates. *Crit. Rev. Food Sci. Nutrit.* **2017**, *57*, 1119–1132. [CrossRef]

9. de Castro, R.J.S.; Bagagli, M.P.; Sato, H.H. Improving the functional properties of milk proteins: Focus on the specificities of proteolytic enzymes. *Curr. Opin. Food Sci.* **2015**, *1*, 64–69. [CrossRef]

10. Lucas, A.; Sodini, I.; Monnet, C.; Jolivet, P.; Corrieu, G. Probiotic cell counts and acidification in fermented milks supplemented with milk protein hydrolysates. *Intern. Dairy, J.* **2004**, *14*, 47–53. [CrossRef]

11. Sodini, I.; Lucas, A.; Tissier, J.P.; Corrieu, G. Physical properties and microstructure of yoghurts supplemented with milk protein hydrolysates. *Intern. Dairy J.* **2005**, *15*, 29–35. [CrossRef]

12. Chatterjee, A.; Kanawjia, S.K.; Khetra, Y. Properties of sweetened Indian yogurt (mishti dohi) as affected by added tryptic whey protein hydrolysate. *J. Food Sci. Technol.* **2016**, *53*, 824–831. [CrossRef] [PubMed]

13. Raynal, K.; Remeuf, F. The effect of heating on physicochemical and renneting properties of milk: A comparison between caprine, ovine and bovine milk. *Intern. Dairy J.* **1998**, *8*, 695–706. [CrossRef]

14. Raynal-Ljutovac, K.; Park, Y.W.; Gaucheron, F.; Bouhallab, S. Heat stability and enzymatic modifications of goat and sheep milk. *Small Ruminant Res.* **2007**, *68*, 207–220. [CrossRef]

15. López-Fandiño, R. High pressure-induced changes in milk proteins and possible applications in dairy technology, a review. *Intern. Dairy J.* **2006**, *16*, 1119–1131. [CrossRef]

16. Devi, A.F.; Buckow, R.; Kasapis, S. Structuring dairy systems through high pressure processing. *J. Food Engin.* **2013**, *114*, 106–122. [CrossRef]

17. Loveday, S.M.; Sarkar, A.; Singh, H. Innovative yoghurts: Novel processing technologies for improving acid milk gel texture- Review. *Trends Food Sci. Technol.* **2013**, *33*, 5–20. [CrossRef]

18. López-Fandiño, R.; De La Fuente, M.; Ramos, M.; Olano, A. Distribution of minerals and proteins between the soluble and colloidal phases of pressurized milks from different species. *J. Dairy Res.* **1998**, *65*, 69–78. [CrossRef]

19. Ferragut, V.; Martinez, V.M.; Trujillo, A.-J.; Guamis, B. Properties of yogurt made from whole ewe's milk treated by high hydrostatic pressure. *Milchwissenschaft* **2000**, *55*, 267–269.

20. Huppertz, T.; Kelly, A.L.; Fox, P.F. High pressure induced changes in ovine milk. Effects on the mineral balance and pH. *Milchwissenschaft* **2006**, *61*, 285–288.

21. Huppertz, T.; Kelly, A.L.; Fox, P.F. High pressure induced changes in ovine milk. Effects on casein micelles and whey proteins. *Milchwissenschaft* **2006**, *61*, 394–397.

22. Moatsou, G.; Katsaros, G.; Bakopanos, C.; Kandarakis, I.; Taoukis, P.; Politis, I. Effect of high-pressure treatment at various temperatures on activity of indigenous proteolytic enzymes and denaturation of whey proteins in ovine milk. *Intern. Dairy J.* **2008**, *18*, 1119–1125. [CrossRef]

23. Bakopanos, C.; Moatsou, G.; Kandarakis, I.; Taoukis, P.; Politis, I. Effect of high pressure treatment at various temperatures on the rennet clotting behavior of bovine and ovine milk. *Milchwissenschaft* **2010**, *65*, 266–269.

24. Moschopoulou, E.; Moatsou, G. Greek dairy products: Composition and processing. In *Mediterranean Food: Composition and Processing*; da Cruz, R.M.S., Vieira, M.M.C., Eds.; Taylor & Francis Group, CRC Press: Boca Raton, FL, USA, 2016; pp. 268–321.

25. Lucey, J.A.; Tamehana, M.; Singh, H.; Munro, P.A. Effect of the interaction between denatured whey proteins and caseins micelles on the formation and the rheological properties of acid skim milk gels. *J. Dairy Res.* **1998**, *65*, 555–567. [CrossRef]

26. Anema, S.G.; Li, Y. Effect of pH on the association of denatured whey proteins with casein micelles in heated reconstituted skim milk. *J. Agric. Food Chem.* **2003**, *51*, 1640–1646. [CrossRef]

27. Kehoe, J.J.; Foegeding, E.A. Interaction between β-Casein and whey proteins as a function of pH and salt concentration. *J. Agric. Food Chem.* **2011**, *59*, 349–355. [CrossRef]

28. Ozcan-Yilsay, T.; Lee, W.-L.; Horne, D.; Lucey, J.A. Effect of trisodium citrate on rheological and physical properties and microstructure of yogurt. *J. Dairy Sci.* **2007**, *90*, 1644–1652. [CrossRef]

29. Jørgensen, C.E.; Abrahamsen, R.K.; Rukke, E.-O.; Hoffmann, T.K.; Johansen, A.-G.; Skeie, S.B. Processing of high-protein yoghurt—A review. *Intern. Dairy, J.* **2019**, *88*, 42–59. [CrossRef]

30. Chua, D.; Deeth, H.C.; Oh, H.E.; Bansal, N. Altering the casein to whey protein ratio to enhance structural characteristics and release of major yoghurt volatile aroma compounds of non-fat stirred yoghurts. *Intern. Dairy J.* **2017**, *74*, 63–73. [CrossRef]

31. Needs, E.C.; Capellas, M.; Bland, A.P.; Manoj, P.; MacDougal, D.; Paul, G. Comparison of heat and pressure treatments of skim milk, fortified with whey concentrate, for set yogurt preparation: Effects on proteins milk and gel structure. *J. Dairy Res.* **2000**, *67*, 329–348. [CrossRef]

32. Walsh-O'Grady, C.D.; O'Kennedy, B.T.; Fitzgerald, R.J.; Lane, C.N. A rheological study of acid-set "simulated yogurt milk" gels prepared from heat- or pressure-treated milk proteins. *Lait* **2001**, *81*, 637–650. [CrossRef]

33. Udabage, P.; Augustin, M.A.; Versteeg, C.; Puvanenthiran, A.; Yoo, J.A.; Allen, N.; McKinnon, I.; Smiddy, M.; Kelly, A.L. Properties of low-fat stirred yoghurts made from high-pressure-processed skim milk. *Innov. Food Sci. Emerg. Technol.* **2010**, *11*, 32–38. [CrossRef]

34. ISO 7889/IDF 117. *Yogurt-Enumeration of Characteristic Microorganisms-Colony-Count Technique at 37 Degrees C*; International Dairy Federation: Brussels, Belgium, 2003.

35. Kyriacou, A.; Tsimpidi, E.; Kazantzi, E.; Mitsou, E.; Kirtzalidou, E.; Oikonomou, Y.; Gatzis, G.; Kotsou, M. Microbial content and antibiotic susceptibility of bacterial isolates from yoghurts. *Intern. J. Food Sci. Nutrit.* **2008**, *59*, 512–525. [CrossRef]

36. Madureira, A.R.; Tavares, T.; Gomes, A.M.P.; Pintado, M.E.; Malcata, F.X. Invited review: Physiological properties of bioactive peptides obtained from whey proteins. *J. Dairy Sci.* **2010**, *93*, 437–455. [CrossRef]

Communication

Characteristics of Instrumental Methods to Describe and Assess the Recrystallization Process in Ice Cream Systems

Anna Kamińska-Dwórznicka [1],*, Ewa Gondek [1], Sylwia Łaba [2], Ewa Jakubczyk [1] and Katarzyna Samborska [1]

[1] Department of Food Engineering and Process Management, Faculty of Food Sciences, Warsaw University of Life Sciences (WULS-SGGW), Nowoursynowska 159C, 02-776 Warsaw, Poland; ewa_gondek@sggw.pl (E.G.); ewa_jakubczyk@sggw.pl (E.J.); katarzyna_samborska@sggw.pl (K.S.)

[2] Institute of Environmental Protection-National Research Institute, Krucza 5/11d St., 00-548 Warsaw, Poland; sylwia.laba@ios.gov.pl

* Correspondence: anna_kaminska1@sggw.pl; Tel.: +48-22-59-37-569; Fax: +48-22-59-37-576

Received: 14 February 2019; Accepted: 1 April 2019; Published: 4 April 2019

Abstract: Methods of testing and describing the recrystallization process in ice cream systems were characterized. The scope of this study included a description of the recrystallization process and a description and comparison of the following methods: microscopy and image analysis, focused beam reflectance measurement (FBRM), oscillation thermo-rheometry (OTR), nuclear magnetic resonance (NMR), splat-cooling assay, and X-ray microtomography (micro-CT). All the methods presented were suitable for characterization of the recrystallization process, although they provide different types of information, and they should be individually matched to the characteristics of the tested product.

Keywords: recrystallization; food hydrocolloids; methods for crystal structure evaluation

---

1. Introduction

The most important factor that determines frozen food quality is the course of crystallization [1]. Crystallization is a process of ice crystal formation as a consequence of atomic ordering and mostly includes hexagonal columns, plates, and dendritic crystal lattices [2,3]. Size, location, and morphology of the ice crystals determine the quality of frozen food, especially ice cream desserts [3–5]. Large ice crystals have a negative impact on product textural properties. Ice crystals of sizes between 10 and 20 µm give the product its desired texture whereas ice crystals larger than 50 µm (if present in a significant quantity) cause the product to have an undesirable (coarse or grainy) texture [1,6–10].

Ice cream is a multiphase physicochemical system originating from the dispersion of individual components in different phases [11,12]. The structure of ice cream is formed by dispersion of air in the frozen liquid that consists of approximately two-thirds water. Therefore, ice cream is a foam, a system in which a liquid (dispersing) phase is dispersed in air (dispersed phase). In the water phase, ice cream is a real solution of sucrose, lactose, and other sugars as well as mineral salts, whose particle sizes do not exceed 1 µm [11]. In general, an ice cream system is constituted of four phases [13]: unfrozen matrix (a solution of different mono- and polysaccharides), air bubbles (with sizes between 20 and 150 µm), ice crystals (with sizes from 10 to 75 µm), and fat globules (between 0.4 and 4 µm). Ice formation occurs after initial freezing, accelerates within the first hours after production, and under unstable temperature conditions, during storage, ice crystals grow due to the recrystallization process [8,14]. Various factors, that include total solids, initial freezing temperature, unfrozen water, stabilizer type, sweetener type, and storage temperature influence the excessive crystal growth during storage [14,15]. When temperature fluctuates, unfrozen water diffuses to the surface of existing crystals and enhances their growth (Figure 1).

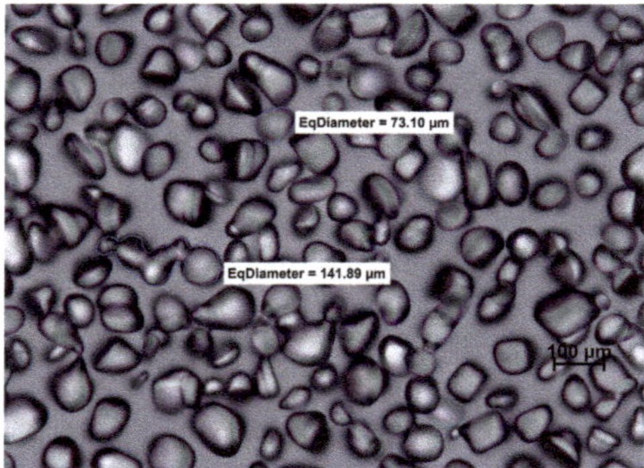

Figure 1. Microscopic image of ice crystals in model sucrose solution (50%) after 96 h of storage at −8 °C; coalescence visible (own work, not published).

The recrystallization process occurs at a constant temperature during long storage, especially above the glass transition temperature [16–18]. Heat and mass transfer cause some crystals to melt and others to grow [19]. During storage, this ice crystal growth occurs mostly because of two mechanisms, coalescence and migration. Coalescence is the process of gathering two or more adjacent ice crystals that form a kind of bridge between them until a single and much larger ice crystal arises. Migration (Ostwald ripening) consists of two stages: melting of smaller crystals, and movement of melted liquid to the surface of crystals with larger diameters. Water molecules at the surface of small crystals are not firmly bound because of the high curvature. These "free" water molecules tend to diffuse through the freeze-concentrated matrix and are deposited on the surface of the crystals with a larger diameter. The water molecule diffusion process occurs because of the differences in vapor pressure (the vapor pressure is inversely proportional to the ice crystal radius). Usually, these two mechanisms of recrystallization occur simultaneously. Some researchers have claimed that the rate of crystal growth may be dependent on the viscosity of the unfrozen phase [6,8,11,14,20]. However, the influence of selected stabilizers on the recrystallization rate in frozen food systems has been investigated most intensively [1,8,21–23].

Hydrocolloid stabilizers are used in food production to modify water-binding capacity, freezing rates, ice crystal formation, and rheological properties [7,8,11,12,24]. Many studies have suggested that some aspects of stabilizer functionality with respect to recrystallization protection may depend on the structure, as measured by rheological properties, which results from the freeze-concentration of the polysaccharide in the unfrozen phase of ice cream. This structure from the stabilizers would affect the rate at which water diffuses to the surface of a growing ice crystal. The stabilizers could also lead to the formation of small curvatures with different radii during ice crystal growth. These newly formed curvatures appear on the surfaces of both smaller and larger ice crystals and prevent differences in vapor pressure between them [1,8,11,21–23]. Polysaccharide stabilizers such as guar gum, locust bean gum (LBG), carboxyl methylcellulose, alginate, and xanthan gum are used commonly to control crystal lattice creation.

Different forms of carrageenan are commonly used as stabilizers. The kappa carrageenan form is mostly used to stabilize dairy products, but it may also be applied to control crystal growth in sorbet production [1,25]. The iota fraction of carrageenan reacts electrostatically with milk proteins to form a three-dimensional network that resists separation of the suspended phase in ice cream mixes [11,26]. Gaukel et al. (2014) [8] investigated the impact of a special protein called antifreeze protein (AFP) on the ice recrystallization inhibition process. Due to the fact that recrystallization is a significant problem in frozen food, recrystallization and its inhibition have both been widely studied. Moreover, currently there is interest in the possibilities of applying different methods to describe and control this process during storage as well. Most of the studies related to the measurement of the recrystallization rate consist of determining the ice crystal size distribution and the ice crystal size using microscopy. It is the best known, although not the only method, to describe the recrystallization phenomenon.

The aim of this review was to outline the basic characteristics of the measurement method, sample preparation, and equipment required to show and describe ice crystals during and after the recrystallization process using the following methods: FBRM, OTR, NMR, splat cooling, microscopy analysis, and X-ray microtomography.

2. Methods of Testing and Describing the Recrystallization Process

2.1. Focused-Beam Reflectance Measurement (FBRM) Technique

Control of the ice crystallization process is mostly conducted in an empirical way, mainly due to a lack of experimental data. Research on ice crystal size distribution (CSD) is not simple, especially because of the possibility of melting under unstable conditions and also when it is about a type of ice cream product that contains three phases [6]. The focused-beam reflectance measurement (FBRM) is a new tool for on-line measurements created to investigate and to monitor CSD during the laboratory and the industrial crystallization processes [6,7,14,27].

The real-time particle size analysis technique of FBRM works by focusing a laser beam directly down the probe tip through a sapphire window. The optical part is rotated about an axis to the probe (2 m/s), so that the beam traces out the circular path (reflected light is detected in the probe). The probe tip is inserted, at an angle, directly into the process streams, to ensure that particles can flow easily across the probe window where the measurement takes place (Figure 2). The laser beam can scan across particle passes near the window. It notes the duration of the reflection and deduces the length of the chord [6,28–30].

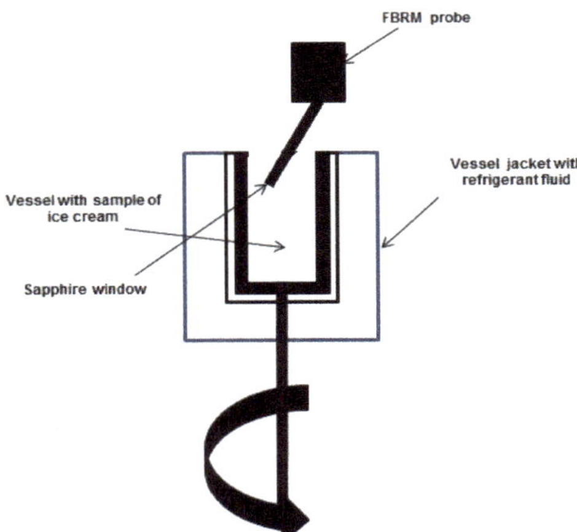

Figure 2. FBRM experimental device (own work, based on Amamou et al., 2010 studies [6]).

Generally, the FBRM instrument can acquire thousands of chord lengths per second. On a counter board, these lengths are classified into a series of size ranges that are expressed as a distribution, referred to as a chord length distribution (CLD). This technique considers the shape and dimensions of the particles. However, the CLD value does not gives information about morphology of the particles and it is less useful for the characterization of a crystal lattice in frozen products [6,7,27].

Amamou et al. (2010) [6] presented a study which examined the freezing step that occurs in a scraped-surface heat exchanger during the manufacture of sorbet. The aim of this investigation (using FBRM technique) was to follow the evolution of ice crystals during the freezing of sorbet in the exchanger and to relate this evolution to process parameters. The measurement showed that this method could be used to follow crystal structure in a sorbet consisting of up to 30% ice and that a decrease of temperature during refrigeration accelerates ice crystallization and favors the formation of smaller crystals. They demonstrated that when the initial sucrose concentration in the solution increased, the ice fraction increased more slowly and the mean chord length was smaller.

Arellano et al. (2012) [27] demonstrated, using an example of sorbet freezing, that the FBRM sensor may be a promising tool for monitoring on-line development of ice crystals in a product containing up to 40% ice. Using the FBRM method they proved that an increase of dasher speed slightly decreases chord length of the ice crystals, due to the higher shear of the product, which leads to the attrition of ice crystals, producing new, smaller ice nuclei via secondary nucleation.

The recrystallization process in ice cream using the FBRM technique was investigated by Ndoye and Alvarez (2015) [14]. They compared two commercial and differently stabilized ice creams using an original and real-time particle counting and sizing method. They stored the ice creams for 154 days at four different temperatures (-5, -8, -12 and -18 °C) and three amplitudes of temperature fluctuations (±0.1, ±0.75, ±2.5 °C). The crystal size distributions (CSD) were assessed at various time intervals and the recrystallization kinetic data were obtained by fitting the experimental results to the asymptotic Ostwald ripening model. As they expected, recrystallization rates increased with mean storage temperature and amplitude of temperature fluctuations. In the samples of ice cream, they compared which of the stabilizing systems worked better, concluding that the carrageenan seemed to be more effective than LBG. For both ice cream samples, it was proven that ice crystal size increased as a function of time.

The main advantage of the FBRM technique seems to be its suitability for on-line measurements of high solid-concentration suspensions and for following rapid crystallization kinetics. However, it only provides information about the total number of ice crystals and the changes of diameters of ice crystals, and therefore the shape and the changes of ice crystal location are measured.

2.2. Oscillatory Thermo-Rheometry Technique (OTR)

It was shown that the dynamics of rheological measurements can be used for characterization of textural properties and structure of foamed dairy emulsion. The dynamic storage modulus G′ (elastic response) and loss modulus G″ (viscous behavior) can provide information about properties of viscoelastic materials.

Stanley et al. (1996) [31] proved that the modulus (G′) greatly increases when the ice cream temperature decreases, hence increasing the ice fraction in ice cream. Smith et al. (2000) demonstrated that microstructure in whipped cream influenced the dynamics of oscillatory storage (G′) and loss (G″) moduli. The parameters both decreased with a coarser foam structure due to increased air bubble sizes during storage. This method, called oscillatory thermo-rheometry (OTR), allows one to distinguish whether the recrystallization process occurred, but does not provide information about the sizes of ice crystals or the changes in shape and location. That kind of investigation is usually complemented by microscopic analysis or the FBRM technique.

Wildmoser et al. (2004) [32] used rheology for the microstructural and sensorial assessment of ice cream samples, produced with application of different ice cream mix compositions and processes. Rheological properties of ice cream were examined in a rotational rheometer (plate-plate geometry). The creaminess and other sensory factors were investigated in order to correlate them with the results gained in the rheometer. This tool was used to perform oscillatory measurements at low deformation amplitudes for three different temperature ranges to assess the rigidity and "scoopability" of ice cream at a low temperature from -10 to -20 °C. The higher the overrun and the smaller the connectivity of ice crystals were, the smaller were the measured values of moduli G′ and G″. In this study, the OTR technique was accomplished using a cryo-scanning electron microscopy (cryo-SEM) to investigate the ice crystals and the air bubble sizes. As the degree of connectivity of ice crystals increased, the storage and loss moduli at temperatures below -10 °C increased. In the temperature range above 0 °C, air and fat phases played a major role in the rheological behavior, due to the total ice crystal melting processes. The loss modulus G″ increased by a factor of approximately 10 when the air content was increased from 0 to 100%.

Eisner et al. (2005) [33] examined the microstructure of ice cream made using a relatively low viscosity vanilla ice cream mixture, prepared in a freezer with outlet temperature of approximately −5 °C and stored for 2 weeks at −25 °C. Using the OTR method and cryo-scanning electron microscopy (LT-SEM) they found that ice cream foam stability correlated with the sensed creaminess and could be improved with smaller air cells and reduced coalescence. At the temperature range from −20 to −10 °C the microstructure of ice crystals was dominant and the storage modulus G′ decreased while the loss modulus G″ showed a plateau, correlated with the rigidity and "scoopability" of the ice cream. For temperatures from −10 to 0 °C the ice fraction decreased significantly. At temperatures above 0 °C both moduli (G′ and G″) showed a lower plateau, which correlated with the sense of creaminess. Microstructural study may improve fat agglomeration with further enhancement of foam stability, and it correlates with reduced ice crystal sizes.

Sensory quality closely correlates with viscoelastic properties of products. Tsevdou et al. (2015) [34] correlated vanilla ice cream sensory characteristics with the changes in rheological behavior (using the OTR method), during storage under static and dynamic temperature conditions. The formation of ice crystals was estimated in terms of mouthfeel perception. For temperatures in the range of −30 to −5 °C changes in the value of G′ (loss modulus) were observed over a certain period of time, suggesting that the recrystallization phenomenon is not only time, but also temperature dependent. At high storage temperatures such as −5 °C, the G′ value showed that ice recrystallization occurred to a greater extent than at temperatures below −12 °C. It was found that viscoelastic properties correlated with sensory perception for ice crystal formation during storage at isothermal temperature conditions and temperature fluctuations, and thus could be used to predict the quality and the remaining shelf-life of ice cream without recrystallization changes. However, the OTR method did not provide information about changes in shapes and diameters of ice crystals, and therefore it cannot give a precise evaluation of the recrystallization process.

2.3. Nuclear Magnetic Resonance (NMR) Technique

Content, purity, and molecular structure of a sample can be determined using an analytical chemistry technique called nuclear magnetic resonance (NMR). When the NMR technique is used it is possible to quantitatively analyze mixtures containing known compounds. For unknown compounds, NMR can either be used to match against spectral libraries or to infer the basic structure directly. The NMR technique can be used to determine molecular conformation in solution and to study physical properties at the molecular level such as conformational exchange, phase changes, solubility, and diffusion. In order to achieve the desired results a variety of NMR techniques are available. In addition, NMR is versatile and has the potential to be nondestructive, which makes it a potential tool in quality control of various products, including milk-based desserts [35].

The principle of the method relates to the fact that many nuclei possess spin and all nuclei are electrically charged. If an external magnetic field is applied, energy transfer is possible from the base energy to a higher energy level (generally a single energy gap). The energy transfer takes place at a wavelength that corresponds to the radiofrequencies, and when the spin returns to its base level, energy is emitted at the same frequency. The signal that matches this transfer is measured in many ways and processed in order to obtain an NMR spectrum for the charged nucleus [36,37].

The NMR technique has been shown to be an appropriate method to calculate the amount of unfrozen water in a food sample. Lucas et al. (2004) [38] examined liquid from the solid water in aqueous sucrose solutions (sucrose and/or casein). They considered spin-spin relaxation measurements that were usually used and also spin-lattice ones. They showed that spin-lattice relaxation provides information about the ice molecular structure. This work confirmed that the ice phase in the case of sucrose solutions is composed of pure water.

Hagiwara et al. (2006) [39] investigated the relationship between the recrystallization rate of ice crystals in sugar solutions (sucrose, maltose, glucose and fructose) and the water mobility in a freeze-concentrated matrix. They observed ice crystals during the recrystallization process using the cryomicroscope system. Their study was complemented with an NMR study in order to examine water mobility via the self-diffusion coefficient of the water component. They found that the recrystallization rate in a variety of sugar solutions depended significantly on the water mobility in the freeze-concentrated matrix and that the self-diffusion coefficient of the water component was a useful parameter to predict and control the recrystallization rate. Brown et al. (2014) [40] used NMR relaxation and time-dependent self-diffusion measurements to monitor the three-dimensional changes to the vein network in ices with and without addition of the ice binding proteins (IBP called antifreeze protein and AFP). They found that the NMR technique was useful in evaluation of the impact of IBPs (among other things) on the vein network structure and the recrystallization process. The IBPs were found useful to inhibit recrystallization and to modify the three-dimensional ice structures, resulting in persistent small size of ice crystals and shorter diffusion of distances along the vein.

It has been discovered that NMR could be used to determine fat globule size in ice cream and to determine the effect of the formulation on hard ice creams' structure [36,41,42]. Lucas et al. (2005) [43] presented the NMR technique as a nondestructive method to characterize the behavior of both fat and water in ice cream mixtures in the frozen state. They proved that the NMR technique described the crystallized and liquid phases separately, and that they could be applied to determine the amount of unfreezable water and mobility of the freeze-concentrated phase. The NMR technique was also used to determine the impact of the quantity of crystals and their organization on the mechanical properties and textures of ice cream mixes [35,43].

The NMR technique does not involve any thermal processes to assess the amount of ice and thus can be performed at a stable temperature. It is also important that the relaxation parameters of water were used to provide information on the water/non-water molecule interactions. This technique may be a valuable tool for understanding how various stabilizers affect the three-dimensional vein network and recrystallization processes.

2.4. Splat-Cooling Assay

In order to determine the ice recrystallization inhibition (IRI) activity of ice cream stabilizers, the capillary method or splat-cooling assay can be performed [44]. Although it is an older method, it is a simple and valuable capillary method for studying recrystallization inhibition which is usually based on loading the samples into 10 µL glass capillaries (51 mm long, 1 mm outer diameter) by capillary action. Capillaries containing dilution series (for example different AFP concentrations to determine at what level that IRI activity is lost) are later folded together exactly, with no space between adjacent capillaries. Subsequently, all the series were snap frozen in different organic compounds, for example, 2,2,4-trimethylpentane or 95% ethanol and cooled to approximately −60°C with dry ice. After the snap freezing, samples were immersed in a jacketed beaker filled with 50% ethylene glycol in order to maintain the temperature at the level of −6 °C. The incubation took 16 to 20 h and later microscopy and image capture were done using polarizing light filters [45,46]. These methods used a simple set-up and allowed the analysis of the IRI activity for a series of samples within one field of view. Tomczak et al. (2003) [45] claimed that the capillary method allowed samples to be aligned and viewed simultaneously, which facilitated the determination of the IRI endpoint. They noted that after the samples have been prepared they could be archived in a freezer for future IRI activity analysis. In fact, sample preparation was not so easy and could be problematic.

The IRI activity of different substances such as AFP and some polysaccharides can be demonstrated using "splat" assay. This is not a new method and has been used for nearly twenty years. In these experiments, a small volume of sample was usually expelled from a height of 1.5 to 3 m onto a metal plate (for example an aluminum block) that had been cooled in liquid nitrogen or dry ice. The sample drop froze upon hitting the metal plate, forming a thin (splat) wafer of ice. This wafer was later transferred to a microscope stage at a high sub-zero temperature where the sample recrystallized over time and annealing temperature (for example: −10, −8, −6, −4 °C). Different modifications of this method are possible, for example, the sample could be snap frozen between two cover slips. This technique was complemented by crossed polarizers with a dissecting microscope and ice-binding activities of some substances that were examined by Raymond and Knight (2003) [47]. The splat cooling technique had the advantage that ice crystal growth was directly observed and hence it was easier to interpret. However, preparation of each individual coverslip was problematic, and in addition must be photographed separately and only after the photographs or images are assembled and analyzed [45,46,48].

2.5. Microscopy and Image Analysis

The most popular and universal method of testing and describing the recrystallization process is direct light microscopy observation followed by image analysis of the ice crystals. When the work was started with another technique such as oscillatory thermos-rheometry or splat cooling assay, to complete it the work was usually supported by an appropriate microscopy technique, a cryo-scanning electron microscope or a microscope with a polarizer [32,44]. Physicists studying polar ice structures proposed examining the structures by the method of direct observation using an optical microscopy with episcopic coaxial lighting [49], and it was later adapted by Faydi et al. (2001) [50]. Caillet et al. (2003) [51] recommended the direct microscopy method as a good technique for analysis of frozen food structure. They compared it with two other methods: a destructive method by dispersion and observation by light microscopy; and an indirect method, by scanning electron microscopy after freeze-drying the sample. The three methods examined led to the same conclusions for the examination under the same conditions of freezing.

Donhowe and Hartel (1996a, 1996b) [16,17] examined sizes of ice crystals in ice cream with an optical microscope placed in a refrigerated glove box. Samples of ice cream were stored at −14 °C for several hours. Photomicrographs of ice crystals were taken within 15 min of the sample preparation (no change in ice crystal size occurred over the time period of measurement). Negatives were enlarged and analyzed. This assay led to the conclusion that recrystallization in ice cream stored in bulk containers increased with mean storage temperature for both constant and varying temperatures. Ice crystal sizes in these samples increased linearly with time as well.

Regand and Goff (2003) [12] presented a study of stabilized ice cream model systems. Small drops of different solutions with the addition of some hydrocolloid stabilizers were placed between a slide and cover slips, frozen to −50 °C, then cycled between −3.5 °C and −6 °C on a cold stage of the light microscope, and then the images were acquired using a camera. The ice crystals were later counted and measured individually from the images (at least 200 crystals for the sample). They based the conclusion on a logistic model of ice crystal size distributions characterized earlier by Flores and Goff (1999a, 1999b) [52,53]. This method allowed them to obtain the ice crystal diameter at 50% of cumulative distribution of the sample (X_{50}) and the slope of cumulative distribution at X_{50}. The recrystallization rate was calculated as the slope of linear regression of the curve plotted with values of X_{50} for each cycle. Using this method, they proved there was significant retardation of recrystallization with the addition of sodium alginate and xanthan. The same technique (using different equipment and a different image analysis program) was first used to explain the IRI activity of kappa carrageenan hydrolysates in model sucrose solutions [4] (Table 1) and later in ice cream sorbet [1] (Table 2).

Table 1. X_{50} parameter value for model sucrose solutions with the addition of kappa carrageenan and its hydrolysates (value estimated from data presented by Kamińska-Dwórznicka et al., 2015 [4]).

Sample	X_{50} (after 24 h) (μm)	X_{50} (after 96 h) (μm)
30% suc + KK	6	20
30% suc + 3 h HCL	7	8
30% suc + 1.5 h H_2SO_4	7	12

Explanatory notes: suc: sucrose solutions, KK: kappa carrageenan, 3 h HCL: hydrolysates after 3 h of hydrolysis in HCL acid, 1.5 h H_2SO_4: hydrolysates after 1.5 h of hydrolysis in H_2SO_4 acid.

Table 2. X_{50} parameter value for strawberry sorbet with the addition of kappa carrageenan and its hydrolysates (value estimated from data presented by Kamińska-Dwórznicka et al., 2015 [1]).

Sample	X_{50} (after 24 h) (μm)	X_{50} (after 96 h) (μm)
sorbet + KK	6	20
sorbet + 3 h HCL	7	8
sorbet +1.5 h H_2SO_4	7	12

Explanatory notes: KK: kappa carrageenan, 3 h HCL: hydrolysates after 3 h of hydrolysis in HCL acid, 1.5 h H_2SO_4: hydrolysates after 1.5 h of hydrolysis in H_2SO_4 acid.

The method described above makes it possible to analyze not only the changes in diameter of ice crystals but also their shape and location. On the basis of the images it is possible to analyze the mechanism of recrystallization. From the shape of the ice crystal, it is sometimes easy to read that accretion between adjacent crystals occurred (Figure 1). Numerous studies have shown that the shape of the ice crystal is influenced by the temperature cycle and by the addition of active substances. Gaukel et al. (2014) [8] focused on crystal morphology in model sucrose solutions with the addition of different types of AFPs. They constructed their theory based on microscope and image analysis. On the basis of the shape of the ice crystals, they claimed that the kappa carrageenan molecule interacted with the ice crystal surface similar to the AFP interactions. The AFPs (also called IBPs) were identified in the blood of Antarctic fish, and then they were found in different organisms. It was discovered [54] that by adsorption of AFP to the ice crystal surface, ice growth was only possible between proteins, leading to a micro curvature. The exact IRI mechanism of AFP is still not fully understood. In the case of pure sucrose solution (Figure 1), typical round ice crystals were formed. In contrast (Figure 3), the shape of ice crystals in sucrose with AFP III present was angular, elongated, and gave the illusion of a three-dimensional structure.

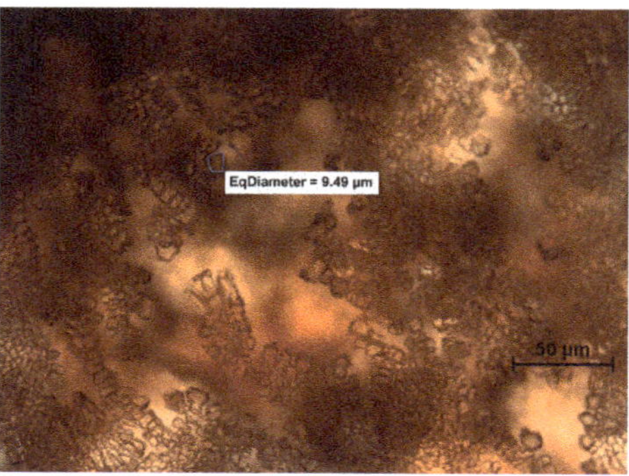

Figure 3. Microscopic image of ice crystals in strawberry sorbet with the addition of AFP III (0.000002%) after one month of storage at −18 °C (own work, not published).

When the growth rate of ice crystals in samples with the IRI active substances is small, there is no difference in morphology or size of ice crystals, as was observed by Kamińska-Dwórznicka et al. (2016) [5] for kappa carrageenan hydrolysates added to model sucrose solutions (Figure 4).

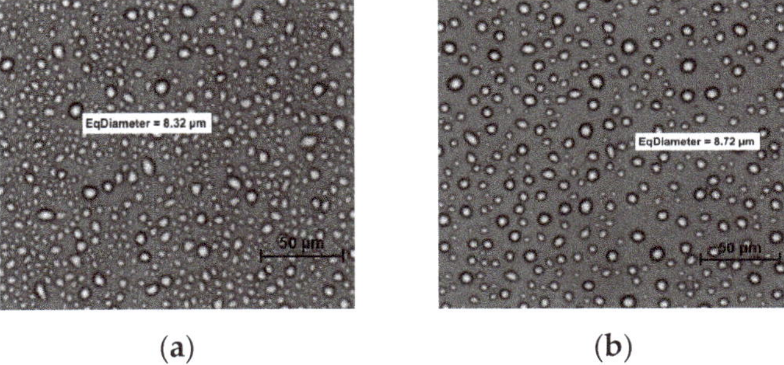

(**a**)   (**b**)

Figure 4. Microscopic images of ice crystals in model sucrose solutions with the addition of enzymatic hydrolysates of κ-carrageenan (after HCL hydrolysis), after 24 h (**a**), and 96 h (**b**) of storage at −8 °C (Kamińska-Dwórznicka et al., 2016 [5]).

2.6. X-ray Microtomography (Micro-CT)

The technique of microscopy and image analysis has brought positive results in ice crystal microstructure investigations. However, it is a two-dimensional technique and it does not characterize multidimensional structure of ice cream mix before and after freezing. During the recrystallization process, the microstructure of ice cream (Figure 5) changed (not only in ice crystal size but also in size of fat globules and air bubbles). It was possible to overcome these (and other limitations) with the X-ray microtomography (micro-CT) three-dimensional technique [55,56].

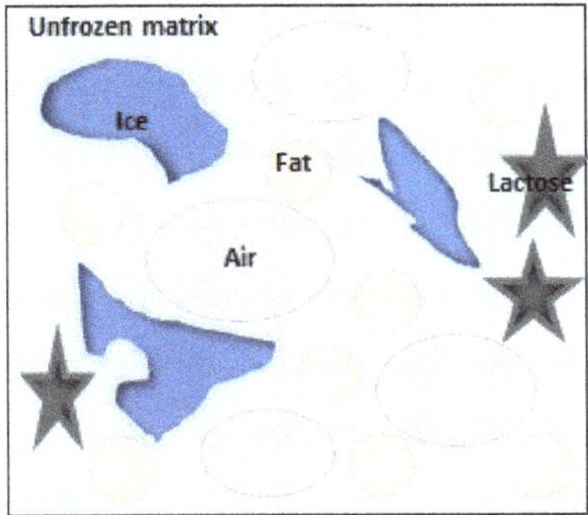

Figure 5. Schematic diagram showing the complex microstructure of ice cream visible during micro-CT analysis (own work, based on Guo et al., 2017 [57]).

The micro-CT method is based on the same assumptions as classic tomography. However, by using a smaller radiation spot, it is possible to obtain a higher resolution of the reconstructed image [55,58]. This non-destructive method was used by Pinzer et al. (2012) [13] to examine the three-dimensional distribution of the three main phases in ice cream model systems. A cylindrical sample holder (10 mm diameter) was filled with a piece of ice cream and placed inside the CT scanner which was programmed to start a scan every 4 h. The cold laboratory was programmed to follow a temperature step function, varying between −20 and −8 °C. When analyzing only the process of recrystallization, they found that during cold periods elongated ice crystals were formed. Because of the temperature fluctuation, partial melting occurred and elongated ice crystals split up again into smaller ones. They concluded that a partial melting–refreezing mechanism was the dominant coarsening mechanism for the investigated storage conditions. Hence, both the size and structure of ice crystals, as well as the mechanisms of those microstructural changes were examined. However, they encountered some difficulties in the interpretation of images caused by the limited resolution of desktop microtomographs and the poor contrast between different phases, which introduced systematic errors.

Guo et al. (2017) [57] presented results for micro-CT measurements of thermal changes in the microstructure of ice creams after 0, 7 and 14 cycles at temperatures ranging between -15 and $-5\,^\circ$C. For tomography, 3 mm diameter tubes were filled with ice creams just before the analysis (a bed of dry ice was used to avoid any changes). With respect to the recrystallization process, they found that melting and solidification had the greatest impact on the final ice cream microstructure. Temperature conditions during the first seven thermal cycles promoted migratory recrystallization. The sizes of the ice crystals increased while the number of the ice crystals decreased. However, the growth rate of crystals after the seven cycles decreased significantly, which was related to the limited amount of available water from the unfrozen matrix, adjacent ice crystals, and air cells.

3. Conclusions

The technique of microscopy and image analysis allows one to describe ice crystal microstructure. From the images of the ice crystals we can easily obtain information about the size and the changes of their shapes and location during storage at different temperature and time conditions. The images can be easily analyzed using specific computer software. The main disadvantages of this method are difficulties in the preparation of samples and its influence on the repeatability of results. The technique of X-ray microtomography seems to offer a new possibility in the analysis of the recrystallization process as a non-destructive method that shows ice cream samples in 3D, but has some difficulties with the final interpretation of images. The FBRM (focused beam reflectance) technique is fully automated and provides results more easily and faster than simple image microscopy and image analysis. It is a suitable method for in situ measurements, and it allows sample preparation to be avoided because the measurement is conducted by the probe immersed in the ice cream mixture. However, it can only provide information about changes in the diameters of crystals, without shape and location analysis. The OTR (oscillatory thermo-rheometry) technique is a method in which viscoelastic properties of ice cream closely correlate with the sensory quality, and hence it provides information about the shelf-life of ice cream without recrystallization changes. However, the changes in size or shapes and location of ice crystals are not measured. Nuclear magnetic resonance (NMR) is an effective method to evaluate the amount of unfrozen water in a food sample. Hence, it is a valuable tool for understanding the impact of a stabilizer on the recrystallization processes without providing any information about sizes of ice crystals and locations. Splat-cooling assay is the oldest method to describe the amount, sizes, and morphology of ice crystals. It is a method with a very specific technology for sample preparation and is not suitable for different types of frozen food.

All of the discussed methods are suitable for describing the recrystallization processes, although they provide different types of information, and they should be matched individually to the characteristics of the tested product.

Author Contributions: Conceptualization, A.K.-D. and K.S.; software, K.S. and E.J.; validation, E.G. and S.Ł.; investigation and analysis of particular methods: 2.1. A.K.-D. and K.S., 2.2. A.K.-D. and S.Ł., 2.3. E.J. and E.G., 2.4. A.K.-D., and S.Ł., 2.5. A.K.-D., K.S., and E.G., 2.6. A.K.-D., E.G., and E.J.; writing—original draft preparation, A.K.-D.; writing—review and editing, A.K.-D. and K.S.; supervision, K.S., S.., and E.G.

Funding: This research received no external funding.

References

1. Kamińska-Dwórznicka, A.; Matusiak, M.; Samborska, K.; Witrowa-Rajchert, D.; Gondek, E.; Jakubczyk, E.; Antczak, A. The influence of kappa carrageenan and its hydrolysates on the recrystallization process in sorbet. *J. Food Eng.* **2015**, *167*, 162–165. [CrossRef]
2. Delgado, A.E.; Sun, D.-W. Ultrasound-accelerated freezing. In *Handbook of Frozen Food Processing and Packaging*; Sun, D.-W., Ed.; CRC Prezz, Taylor & Francis Group: Boca Raton, FL, USA, 2012; pp. 645–666.
3. Kaale, L.D.; Eikevik, T.M. The development of ice crystals in food products during the superchilling process and following storage, a review. *Trends Food Sci. Technol.* **2014**, *39*, 91–103. [CrossRef]
4. Kamińska-Dwórznicka, A.; Antczak, A.; Samborska, K.; Lenart, A. Acid hydrolysis of kappa-carrageenan as a way of gaining new substances for freezing process modification and protection from excessive recrystallization of ice. *Int. J. Food Sci. Technol.* **2015**, *50*, 1799–1806. [CrossRef]
5. Kamińska-Dwórznicka, A.; Skrzypczak, P.; Gondek, E. Modification of kappa carrageenan by β-galactosidase as a new method to inhibit recrystallization of ice. *Food Hydrocoll.* **2016**, *61*, 31–35. [CrossRef]
6. Amamou, A.H.; Benkhelifa, H.; Alvarez, G.; Flick, D. Study of crystal size evolution by focused-beam reflectance measurement during the freezing of sucrose/water solutions in a scraped-surface heat exchanger. *Proc. Biochem.* **2010**, *45*, 1821–1825. [CrossRef]
7. Arellano, M.; Flick, D.; Benkhelifa, H.; Alvarez, G. Rheological characterisation of sorbet using pipe rheometry during the freezing process. *J. Food Eng.* **2013**, *119*, 385–394. [CrossRef]
8. Gaukel, V.; Leiter, A.; Spiess, W.E.L. Synergism of different fish antifreeze proteins and hydrocolloids on recrystallization inhibition of ice in sucrose solutions. *J. Food Eng.* **2014**, *141*, 44–50. [CrossRef]
9. Kamińska, A.; Gaukel, V. Monitoring the growth of crystals in ice cream. *Food. Sci. Technol. Qual.* **2009**, *62*, 57–64. (In Polish)
10. Smith, A.K.; Goff, H.D.; Kakuda, Y. Microstructure and rheological properties of whipped cream as affected by heat treatment and addition of stabilizer. *Int. Dairy J.* **2000**, *10*, 295–301. [CrossRef]
11. Adapa, S.; Schmidt, K.A.; Jeon, I.J.; Herald, T.J.; Flores, R.A. Mechanisms of ice crystallization and recrystallization in ice cream: A review. *Food Rev. Int.* **2000**, *16*, 259–271. [CrossRef]
12. Regand, A.; Goff, H.D. Structure and ice recrystallization in frozen stabilized ice cream model systems. *Food Hydrocoll.* **2003**, *17*, 95–102. [CrossRef]
13. Pinzer, B.R.; Medebach, A.; Limbach, H.J.; Dubois, C.; Stampanoni, M.; Schneebeli, M. 3D—Characterization of three-phase systems using X-ray tomography: Tracking the microstructural evolution in ice cream. *Soft Matter* **2012**, *8*, 4584–4594. [CrossRef]
14. Ndoye, F.T.; Alvarez, G. Characterization of ice recrystallization in ice cream during storage using the focused beam reflectance measurement. *J. Food Eng.* **2015**, *148*, 24–34. [CrossRef]
15. Sutton, R.L.; Wilcox, J.E.A. Recrystallization in ice cream as affected by stabilizers. *J. Food Sci.* **1998**, *63*, 104–107. [CrossRef]
16. Donhowe, D.P.; Hartel, R.W. Recrystallization of ice in ice cream during controlled accelerated storage. *Int. Dairy J.* **1996**, *6*, 1191–1208. [CrossRef]
17. Donhowe, D.P.; Hartel, R.W. Recrystallization of ice during bulk storage of ice cream. *Int. Dairy J.* **1996**, *6*, 1209–1221. [CrossRef]
18. Hartel, R.W. Ice crystallization during the manufacture of ice cream. *Trends Food Sci. Technol.* **1996**, *7*, 315–321. [CrossRef]
19. Cook, K.L.K.; Hartel, R.W. Mechanisms of ice crystallization in ice cream production. *Compr. Rev. Food Sci. Food Saf.* **2010**, *9*, 213–222. [CrossRef]
20. Goff, H.D.; Caldwell, K.B.; Stanley, D.W.; Maurice, T.J. Influence of polysaccharides on the glass transition in frozen sucrose solutions and ice cream. *J. Dairy Sci.* **1993**, *76*, 1268–1277. [CrossRef]
21. Damodaran, S. Inhibition of ice crystal growth in ice cream mix by gelatin hydrolysate. *J. Agri. Food Chem.* **2007**, *55*, 10918–10923. [CrossRef]
22. Regand, A.; Goff, H.D. Effect of biopolymers on structure and ice recrystallization in dynamically frozen ice cream model system. *J. Dairy Sci.* **2002**, *85*, 2722–2732. [CrossRef]
23. Soukoulis, C.; Tzia, C. Impact of the acidification process, hydrocolloids and protein fortifiers on the physical and sensory properties of frozen yogurt. *Int. J. Dairy Technol.* **2008**, *61*, 170–177. [CrossRef]

24. Rochas, C.; Heyraud, A. Acid and enzymic hydrolysis of kappa carrageenan. *Polym. Bull.* **1981**, *5*, 81–86. [CrossRef]
25. Rudolph, B. Seaweed products: Red algae of economic significance. In *Marine and Freshwater Products Handbook*; Martin, R.E., Carter, E.P., Flick, G.J., Davis, L.M., Eds.; Technomic Publishing Company, Inc.: Lancaster, PA, USA, 2000; pp. 515–529.
26. Marshall, R.T.; Arbuckle, W.S. *Ice Cream*, 5th ed.; Chapman and Hall: New York, NY, USA, 1996.
27. Arellano, M.; Benkhelifa, H.; Flick, D.; Alvarez, G. Online ice crystal size measurements during sorbet freezing by means of the focused beam reflectance measurement (FBRM) technology. Influence of operating conditions. *J. Food Eng.* **2012**, *113*, 351–359. [CrossRef]
28. Greaves, D.; Boxall, J.; Mulligan, J.; Montesi, A.; Creek, J.; Solan, E.D.; Koh, C.A. Measuring the particle size of a known distribution using the focused beam reflectance measurement technique. *Chem. Eng. Sci.* **2008**, *63*, 5410–5419. [CrossRef]
29. Hukkanen, E.J.; Braatz, R.D. Measurement of particle size distribution in suspension polymerization using in situ laser backscattering. *Sens. Actuators B Chem.* **2003**, *96*, 451–459. [CrossRef]
30. Wynn, E.J.W. Relationship between particle-size and chord-length distributions in focused beam reflectance measurement: Stability of direct inversion and weighting. *Powder Technol.* **2003**, *133*, 125–133. [CrossRef]
31. Stanley, D.W.; Goff, H.D.; Smith, A.K. Texture-structure relationship in foamed dairy emulsions. *Food Res. Int.* **1996**, *29*, 1–13. [CrossRef]
32. Wildmoser, H.; Scheiwiller, J.; Windhab, E.J. Impact of disperse microstructure on rheology and quality aspects of ice cream. *LWT Food Sci. Technol.* **2004**, *37*, 881–891. [CrossRef]
33. Eisner, M.D.; Wildmoser, H.; Windhab, E.J. Air cell microstructuring in a high viscous ice cream matrix. *Colloids Surf. Physicochem. Eng. Asp.* **2005**, *263*, 390–399. [CrossRef]
34. Tsevdou, M.; Gogou, E.; Dermesonluoglu, E.; Taoukis, P. Modelling the effect of storage temperature on the viscoelastic properties and quality of ice cream. *J. Food Eng.* **2015**, *148*, 35–42. [CrossRef]
35. Bahram-Parvar, M. A review of modern instrumental techniques for measurements of ice cream characteristics. *Food Chem.* **2015**, *188*, 625–631. [CrossRef]
36. Barbosa, L.L.; Sad, C.M.S.; Morgan, V.G.; Figueiras, P.R.; Castro, E.R.V. Application of low field NMR as an alternative technique to quantification of total acid number and sulphur content in petroleum from Brazilian reservoirs. *Fuel* **2016**, *176*, 146–152. [CrossRef]
37. Schuff, N. In Vivo NMR methods, overview of techniques. In *Reference Module in Chemistry, Molecular Sciences and Chemical Engineering Encyclopedia of Spectroscopy and Spectrometry*, 3rd ed.; Elsevier: Amsterdam, The Netherlands, 2017; pp. 211–215.
38. Lucas, T.; Mariette, F.; Dominiawsyk, S.; Le Ray, D. Water, ice and sucrose behavior in frozen sucrose-protein solutions as studied by 1H NMR. *Food Chem.* **2004**, *84*, 77–89. [CrossRef]
39. Hagiwara, T.; Hartel, R.W.; Matsukawa, S. Relationship between recrystallization rate of ice crystals in sugar solutions and water mobility in freeze-concentrated matrix. *Food Biophys.* **2006**, *1*, 74–82. [CrossRef]
40. Brown, J.R.; Seymour, J.D.; Brox, T.I.; Skidmore, M.L.; Wang, C.; Christner, B.C.; Luo, B.; Codd, S.L. Recrystallization inhibition in ice due to ice binding protein activity detected by nuclear magnetic resonance. *Biotechnol. Rep.* **2014**, *3*, 60–64. [CrossRef]
41. Gabriele, D.; Migliori, M.; Sanzo, R.D.; Rossi, C.O.; Ruffolo, S.A.; de Cindio, B. Characterisation of dairy emulsions by NMR and rheological techniques. *Food Hydrocoll.* **2009**, *23*, 619–628. [CrossRef]
42. Lucas, T.; Ray, D.L.; Barey, P.; Mariette, F. NMR assessment of ice cream: Effect of formulation on liquid and solid fat. *Int. Dairy J.* **2005**, *15*, 1225–1233. [CrossRef]
43. Lucas, T.; Wagener, M.; Barey, P.; Mariette, F. NMR assessment of mix and ice cream. Effect of formulation on liquid water and ice. *Int. Dairy J.* **2005**, *15*, 1064–1073. [CrossRef]
44. Knight, C.A.; Hallett, J.; DeVries, A.L. Solute effects on ice recrystallization: An assessment technique. *Cryobiology* **2012**, *25*, 55–60. [CrossRef]
45. Tomczak, M.M.; Marshall, C.B.; Gilbert, J.A.; Davies, P.L. A facile method for determining ice recrystallization inhibition by antifreeze proteins. *Biochem. Biophys. Res. Commun.* **2003**, *311*, 1041–1046. [CrossRef]
46. Yu, S.O.; Brown, A.; Middleton, A.J.; Tomczak, M.M.; Walker, V.K.; Davies, P.L. Ice restructuring inhibition activities in antifreeze proteins with distinct differences in thermal hysteresis. *Cryobiology* **2010**, *61*, 327–334. [CrossRef]

47. Raymond, J.A.; Knight, C.A. Ice binding, recrystallization inhibition, and cryoprotective properties of ice-active substances associated with Antarctic sea ice diatoms. *Cryobiology* **2003**, *46*, 174–181. [CrossRef]
48. Wharton, D.A.; Wilson, P.W.; Mutch, J.S.; Marshall, C.J.; Lim, M. Recrystallization inhibition assessed by splat cooling and optical recrystallometry. *CryoLetters* **2007**, *28*, 61–68.
49. Arnaud, L.; Gay, M.; Barnola, J.M.; Duval, P. Imaging of firn and bubbly ice in coaxial reflected light: A new technique for the characterization of these porous media. *J. Glaciol.* **1998**, *44*, 326–332. [CrossRef]
50. Faydi, E.; Andrieu, J.; Laurent, P. Experimental study and modeling of the ice crystal morphology of model standard ice cream. Part I: Direct characterization method and experimental data. *J. Food Eng.* **2001**, *48*, 238–291.
51. Caillet, A.; Cogne, C.; Andrieu, J.; Laurent, P.; Rivoire, A. Characterization of ice cream structure by direct optical microscopy. Influence of freezing parameters. *LWT Food Sci. Technol.* **2003**, *36*, 743–749. [CrossRef]
52. Flores, A.A.; Goff, H.D. Ice crystal size distributions in dynamically frozen model solutions and ice cream as affected by stabilizers. *J. Diary Sci.* **1999**, *82*, 1399–1407. [CrossRef]
53. Flores, A.A.; Goff, H.D. Recrystallization in ice cream after constant and cycling temperature storage conditions as affected by stabilizers. *J. Dairy Sci.* **1999**, *82*, 1408–1415. [CrossRef]
54. Raymond, J.A.; DeVries, A.L. Adsorption inhibition, as a mechanisms of freezing resistance in polar fishes. *Proc. Natl. Acad. Sci. USA* **1977**, *74*, 2589–2593. [CrossRef]
55. Barbetta, A.; Bedini, R.; Pecci, R.; Dentini, M. Role of X-ray microtomography in tissue engineering. *Ann. Ist. Super. Sanita* **2012**, *48*, 10–18.
56. Kareh, K.M.; Lee, P.D.; Gourlay, C.M. In situ, time-resolved tomography for validating models of deformation in semi-solid alloys. *IOP Conf. Ser. Mater. Sci. Eng.* **2012**, *33*, 012037. [CrossRef]
57. Guo, E.; Zeng, G.; Kazantsev, D.; Rockett, P.; Bent, J.; Kirkland, M.; Dalen, G.; Eastwood, D.S.; StJohn, D.; Lee, P.D. Synchrotron X-ray tomographic quantification of microstructural evolution in ice cream—A multiphase soft solid. *RSC Adv.* **2017**, *7*, 15561–15573. [CrossRef]
58. Rockett, P.; Karagadde, S.; Guo, E.; Bent, J.; Hazekamp, J.; Kingsley, M.; Vila-Comamala, J.; Lee, P.D. A 4-D dataset for validation of crystal growth in a complex three-phase material, ice cream. *IOP Conf. Ser. Mater. Sci. Eng.* **2015**, *84*, 012076. [CrossRef]

Review

Production of Milk Phospholipid-Enriched Dairy Ingredients

Zhiguang Huang [1,2], Haotian Zheng [3,4,*], Charles S. Brennan [1,2,5,*], Maneesha S. Mohan [1], Letitia Stipkovits [1], Lingyi Li [5] and Don Kulasiri [1]

[1] Department of Wine, Food and Molecular Biosciences, Faculty of Agriculture and Life Sciences, Lincoln University, Lincoln 7647, Christchurch, New Zealand; Zhiguang.Huang@lincolnuni.ac.nz (Z.H.); Maneesha.Mohan@lincoln.ac.nz (M.S.M.); Letitia.Stipkovits@lincoln.ac.nz (L.S.); Don.Kulasiri@lincoln.ac.nz (D.K.)

[2] Riddet Research Institute, Palmerston North 4442, New Zealand

[3] Department of Food, Bioprocessing and Nutrition Sciences, Southeast Dairy Foods Research Center, North Carolina State University, Raleigh, NC 27695, USA

[4] Dairy Innovation Institute, California Polytechnic State University, San Luis Obispo, CA 93407, USA

[5] Tianjin Key Laboratory of Food and Biotechnology, School of Biotechnology and Food Science, Tianjin University of Commerce, Tianjin 300134, China; lilingyi@tjcu.edu.cn

* Correspondence: haotian.zheng@ncsu.edu (H.Z.); charles.brennan@lincoln.ac.nz (C.S.B.); Tel.: +1-91-9513-2244 (H.Z.); +64-3-423-0637 (C.S.B.)

Received: 27 January 2020; Accepted: 23 February 2020; Published: 2 March 2020

Abstract: Milk phospholipids (MPLs) have been used as ingredients for food fortification, such as bakery products, yogurt, and infant formula, because of their technical and nutritional functionalities. Starting from either buttermilk or beta serum as the original source, this review assessed four typical extraction processes and estimated that the life-cycle carbon footprints (CFs) of MPLs were 87.40, 170.59, 159.07, and 101.05 kg CO_2/kg MPLs for membrane separation process, supercritical fluid extraction (SFE) by CO_2 and dimethyl ether (DME), SFE by DME, and organic solvent extraction, respectively. Regardless of the MPL content of the final products, membrane separation remains the most efficient way to concentrate MPLs, yielding an 11.1–20.0% dry matter purity. Both SFE and solvent extraction processes are effective at purifying MPLs to relatively higher purity (76.8–88.0% *w/w*).

Keywords: milk phospholipids; buttermilk; life-cycle assessment; carbon footprint; supercritical fluid extraction; membrane separation

---

1. Introduction

Milk phospholipids (MPLs) consist of a subclass of polar lipids, namely glycerophospholipids and sphingolipids [1]. Glycerophospholipids comprise a glycerol moiety with two fatty acids esterified at positions *sn*-1 and *sn*-2 and a hydroxyl group at *sn*-3 position, linked to a phosphate group and a polar moiety [1]. The molecular structure of the latter determines the types of glycerophospholipids, namely phosphatidylcholine (PC), phosphatidylserine (PS), phosphatidylethanolamine (PE), phosphatidylinositol (PI), phosphatidyl-glycerol (PG), and phosphatidic acid (PA) [2]. Sphingolipids consist of a sphingosine backbone (2-amino-4-octadecene-1,3-diol) connected to a fatty acid via an amide bond and a polar head. Sphingomyelin (SM), a prominent subclass of sphingolipids, has a phosphocholine residue [1]. In raw bovine milk, the diameters of milk fat globules (MFGs) are around 0.2–15 µm; these MFGs are enveloped by an approximately 15-nm thick tri-layer MFG membrane (MFGM) [3,4]. The composition of MFGM is 30–75% polar lipids, and 25–70% protein, respectively [5]. MPLs lie within the MFGM constructing its backbone. MPLs represent 0.4–1% of the total milk lipids [6], which change with season, lactose stage, and feed [7].

MPLs have exhibited nutraceutical properties due the unique composition of this group of phospholipids. MPLs contain high proportions of SM [8] and PS [9] (24% and 12%, respectively), subclasses which are virtually absent in other sources, such as soy (0% and 0.5%, respectively) and egg yolk lecithin (1.5% and 0%, respectively) [10]. PS is associated with cognitive function and releasing stress, and is replaced by inactive cholesterol as the brain ages [11,12]. SM has been found to be effective in inhibiting colon tumors [13]. Also, MPLs have been implicated in mitigating the risks of Alzheimer's disease and repairing cognitive ability [14], restoring immunological defenses, reducing the incidence of cardiovascular diseases [15,16], and reducing cholesterol absorption and total liver lipids [17]. In addition, MPLs may narrow the gap between formula-fed and breastfed infants concerning neurodevelopment, infectious diseases, and cholesterol metabolism [18,19]. Phospholipid-coated fats, e.g., human breast MFGs, will be properly digested and absorbed, not only due to the size of the MFGs, but also due to the ratio of MFGM proteins to phospholipids [20]. Bovine MPL-enriched ingredients may be used to produce breast milk analogs. For instance, one formula recipe consists of subclasses according to a weight-relative ratio of SM > PC > PE > PS > PI, with 21.1–29.7% SM and 10.2–13.3% PS (both based on total MPLs, similar to those of human breast milk (37.5% and 9%, respectively) [21]. Another infant formula comprises 150 mg/L MPLs [22], mimicking that of breast milk (15–20 mg/dL milk [21] and 0.3–1.0% of the total lipids [23]).

Aside from nutritional value and health benefits, MPLs may provide technical functionalities in food systems, for example, MPLs have been used in the preparation of liposomes [24] and constructing vesicles of bioactive compounds [25]; they are also food emulsifiers and surfactants, foaming agents, texture improvers for bakery goods, and may improve moisture retention for yogurt [26,27].

Many research works and reviews are available on fractionation from buttermilk (BM) and beta serum (BS) [26], isolating MFGs by washing and centrifugation [5], and the membrane separation of polar lipids [8]. However, there is no standard large-scale manufacturing process adopted by the dairy industry. This is due to many reasons. First, the native MFGM is fragile. Shear and turbulent fluid flow can cause damage to the MFGM [28]. These treatments are commonly involved in handling raw milk on farms, in transportation, in silos at manufacturing plants, and during cream separation. Damage to the MFGM may cause associated materials, including MPLs, to deplete from the native MFGs to the aqueous phase of milk. Therefore, more than half of MPLs in raw milk remains in skim milk [29,30]. Second, uncertainties and variables are involved in the MPL fractionation processes. For example, cream washing for removing non-MFGM associated proteins may be performed before butter churning for increased yield or the concentration of MPLs in the resulting BM, or in the retentate of BM after tangential filtration. However, the cream washing procedure may cause a significant change to the MPL composition in BM from unwashed cream [31,32]. Although the mechanism is not clear, it may relate to the physicality of different washing processes. Zheng's group revealed that different washing procedures induce various degrees of damage to MFGM. Therefore, washing may alter the composition of MPL in the fat phase of the washed cream [4,33]. This review aimed to assess different dairy streams rich in MPLs, to evaluate their extraction processes, compare their process intensity and efficiency, and to estimate their life-cycle carbon footprints (CFs) using ISO 14067 and greenhouse gas (GHG) protocols.

2. Milk Phospholipid Extraction from Dairy Products

2.1. Dairy By-Products Rich in Phospholipids

Commercial MPL products are usually derived from dairy products, such as BM [34], BS [8], acid cheese whey BM [35,36], whey protein phospholipids concentrate (WPPC) [37], or whey BM [38]. The dairy streams in Table 1 comprise 2.29%–26.02% MPLs on a dry matter (DM) basis, varying with sources and processes.

BM is the product that remains after the removal of butter by churning cream, which may have been concentrated and/or dried as butter milk powder [39], as illustrated in Figure 1. Acid

BM, a by-product of lactic butter, is made by churning cultured cream. Furthermore, whey BM is produced via the churning of whey cream during cheese making [40]. WPPC is a by-product produced during the microfiltration (MF) of whey for manufacturing whey protein isolate (WPI). The permeate phase (milk-fat-discriminated phase) from this process goes forward for WPI manufacturing and the fat-remaining phase (retentate phase) containing residual whey proteins is further concentrated for producing WPPC. A typical WPPC is comprises more than 12% fat and 50% protein (DM), and less than 8% ash and 6% moisture [37].

BM, the serum phase resulting from the churning of cream, comprises milk proteins and residual fat [34]. In terms of protein, lactose, ash, and DM contents, BS and BM are very similar to those of cream products (Table 1) [41]. For instance, BM (FDC ID 454974) protein content is 3.33%, which is the same as that of cream (FDC ID 495516). Though the fat content of BM is only one-tenth of cream, MPLs of BM are 4–27-fold that of raw milk, as shown in Table 1. The empirical equation MPL = 0.0137 × FC provides an estimation of the MPL content (g/L) of a dairy product, where FC is the fat content of cream [42]. For instance, the estimated BM MPL content of anhydrous milk fat (AMF) from 80% cream, and of butter from 40% cream, is 1.1 and 0.55 g/kg, respectively. Whey BM, a by-product of whey butter, comprises sixfold the MPL content of raw milk, as seen in Table 1 [38].

BM and BS, the most abundant source of MPLs [43], have been underexplored or even treated as a waste stream [44]. For instance, a New Zealand-based dairy manufacturer used two-thirds of their BM for standardization, only one-tenth for BM powder (BMP), and their annual output of MPL concentrate is 320 metric tons [44]. The annual BM output in Canada was 14.1 metric kilotons (18% of butter and 0.5% of bulk liquid [45]), compared to 20 kilotons in Belgium, 16 kilotons in Denmark, and 124.5 kilotons in Germany [46]. In 2013, approximately 5.2 million tons of BM was produced worldwide, similar to that of butter [34]. Worldwide, the annual BMP production was estimated at 410 kilotons (≈9.5% of butter), which has downstream applications for producing ice cream, ingredients-baked foods, low-fat Cheddar cheese, reduced-fat cheese, pizza cheese [40], or in the replacement of skim milk powder for low-fat yogurt [47].

Table 1. Dairy product composition (g/100 g).

Product	Total MPLs	DM MPLs	Fat MPLs	DM Protein	DM Fat	Total Solid	DM Ash	Reference
WPPC	1.60	7.92	29.10	65.00	27.00	20.20	7.92	[48]
WPPC	1.78	1.78	11.63	56.64 ± 0.05	24.23 ± 0.02	97.02	2.57 ± 0.02	[49]
WPPC	2.20	2.20	14.57	64.82 ± 0.12	18.71 ± 0.09	96.40	2.32 ± 0.01	[49]
WPPC	2.20	2.20	14.38	65.00 ± 0.06	18.46 ± 0.01	95.96	2.27 ± 0.03	[49]
BMP	1.30	1.30 ± 0.00	19.01	31.40 ± 0.57	6.84 ± 0.17	-	-	[34]
BM	0.14 ± 0.04	-	-	25.01 ± 0.76	12.22 ± 1.56	-	5.60 ± 0.16	[50]
BM	0.13 ± 0.00	1.43 ± 0.00	25.50	3.46 ± 0.05	0.51 ± 0.02	9.12 ± 0.17	-	[32]
BM	0.16 ± 0.02	1.78 ± 0.17	15.1 ± 0.5	32.44 ± 0.83	11.78 ± 0.53	9.02 ± 0.23	-	[6]
BS	0.97 ± 0.05	8.78 ± 0.41	38.6 ± 2.3	32.41 ± 1.01	22.71 ± 1.04	11.05 ± 0.43	-	[6]
BS	0.93 ± 0.07	8.42 ± 0.63	34.57	3.55 ± 0.11	2.69 ± 0.14	11.05 ± 0.40	-	[8]
BM	0.12 ± 0.01	1.36 ± 0.07	25.36	32.68 ± 0.93	4.87 ± 0.12	8.63 ± 0.26	-	[51]
BS	0.97 ± 0.17	8.8 ± 1.1	40 ± 7	33 ± 3	25 ± 8	11.0 + 0.8	-	[43]
BM	0.11 ± 0.01	1.2 ± 0.1	14 ± 5	33 ± 2	10 ± 5	8.7 ± 0.8	-	[43]
Whey BM	0.16 ± 0.01	2.01 ± 0.16	12.04 ± 0.8	24.89 ± 2.02	16.27 ± 2.06	8.05 ± 0.32	7.01 ± 0.47	[38]
BM454974 [a]	-	-	-	3.33 [b]	3.33 [b]	-	5 [c]	[41]
BM336087 [a]	-	-	-	3.21 [b]	3.31 [b]	12.09	0.69/4.88 [c]	[41]
BM171274 [a]	-	-	-	34.3	5.78	97.03−BMP	7.95	[41]
CM495516 [a]	-	-	-	3.33	36.67	-	3.33 [c]	[41]
CM336519 [a]	-	-	-	2.84	36.08	42.19	2.74 [c]	[41]

MPLs, milk phospholipids; BS, beta serum; BM, buttermilk; BMP, BM powder; WPPC, whey protein phospholipid concentration; CM, cream; DM, dry matter; [a] United States Department of Agriculture (USDA) FoodData Central ID [41]; [b] wet basis; [c] lactose; Ref., reference.

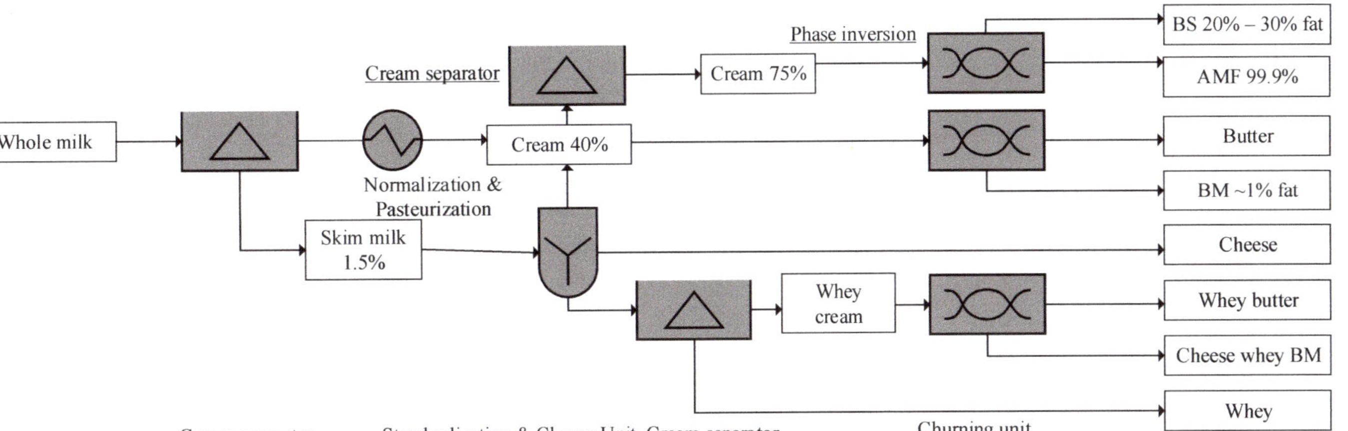

Figure 1. Buttermilk (BM) and beta serum (BS) production process [26]. AMF, anhydrous milk fat. % indicates the fat content.

2.2. Commercialized Milk Phospholipid Products and Concentrate

Phosphoric 500/600/700 and Gangolac 600 (products manufactured by Fonterra) comprise 34%, 75%, 62%, and 15% MPLs, respectively, representing one source of highly-purified MPLs [52,53]. Arla Foods Amba have developed phospholipid-rich, concentrated dairy milk commodities for infant milk formulas and skin care. It has been claimed that Lacprodan® MFGM 10 supports physiological development of the infant gut and provides infants with similar health benefits to breast milk because of their similarities in fatty acid profile [54]. Arla dairy products PL 20/75 consist of 20% and 75% MPLs, respectively [55].

As illustrated in the patents in Table 2, both filtration and solvent extraction are validated processes for manufacturing MPLs. Acetone and supercritical CO_2 are effective solvents for de-fatting. Tatua [56] and Synlait [57] have concentrated MPLs to 5–12.8% (*w/w*, DM basis). Lecico has used membrane separation to produce Lipamine M20 (20% purity) [58].

Table 2. Proprietary/patented manufacturing technologies of milk phospholipids (kg/100 kg products).

Applicant	Input	Technology Used	MPL Content	Reference
Fonterra	BSP	SFE CO_2 defat, hi-pressure DME extract	65.7–75.5	[59]
Meggle	BSP	SFE CO_2 defat, hi-pressure ethanol extract	≈98.5	[60]
Owen John	BSP	SFE CO_2 defat, ethanol co-solvent extract	PI/PS lost	[61]
Arla	BSP	MF, ethanol extraction	16–19	[9]
Merchant & Gould	Cream	UF, DF	27.7–38.8	[62]
Marc Boone	BM	UF 5–20 kDa	≈2.84	[63]
Land O'Lakes	BM, BS	UF, defat using SFE CO_2	>30	[64]
Morinaga	Whey BM	MF 0.2 μm, defat using SFE CO_2	≈22	[65]
Snow Brand	-	Extract using acidic ethanol, defat	-	[66]
Enzymotec	-	Extract using ethanol & hexane, acetone defat	≈24	[21]
Cargill	-	Extract using alcohol (C_1–C_3), acetone defat	-	[67]
Svenska	BMP	Extract using ethanol & n-heptane, acetone defat	≈70 SM	[68]

MPLs, milk phospholipids; BM, buttermilk; BMP, BM powder; BS, beta serum; BSP, BS powder; MF/UF, micro/ultra-filtration; DME, dimethyl ether; PI, phosphatidylinositol; PS, phosphatidylserine; SM; sphingomyelin; SFE, supercritical fluid extraction.

2.3. Laboratory Extraction of Milk Phospholipids

Intact MFGM makes up 2–6% of the total mass of MFG [26]. However, MFGM represents 60%–70% of the total milk phospholipids [69]. Raw bovine milk comprises 0.2–0.4 g MPLs/kg, and raw milk is generally a laboratory source of MPLs [5,70]. Intact MFGs can be isolated with low-speed centrifugation. The cream layer from raw milk skimming can be washed with phosphate buffered saline (PBS; pH 6.8, 0.1 M, 1:10, *v/v*) and centrifuged at 390 g for 10 min at 10 °C. The final cream layer after three washes is the large MFG fraction [71]. Different from isolating intact MFGM, Sanches-Juanes et al. [72] ruptured MFGM and recrystallized milk lipids, and starting from raw milk, they washed cream with a 0.15 M NaCl solution and precipitated casein using centrifugation at 5000× *g*.

Cream washing is a step used to remove casein and other non-MFGM materials from cream [44]. After centrifugation, casein will precipitate, with lipid stratification at the top layer [73]. Also, calcium, naturally present in casein micelles, can form a complex between MFGM and the casein micelles through its binding to the phospho-casein and phospholipids of MFGM, leaving impurities with MPLs [74]. In addition, washing causes a severe loss of phospholipids, almost 60% per wash [32]. Hence, washing facilities for separating MPLs are costly and energy-intensive [44], thereby they are mainly only used for laboratory purposes [5,73,75].

In addition to washing and centrifugation, the microfiltration of raw milk has been applied to produce MFGM material. It has been found that a 1.4-μm ceramic membrane was superior to 0.8 μm, yielding a high-purity MFGM material, which was composed of 7% phospholipids and 30% protein [76].

For analysis purposes, MPL samples are usually prepared using solvent extraction. The Folch [77] and Bligh [78] methods use chloroform and methanol to dissolve lipids. Other lipophilic extraction formulas include the Mojonnier solvents [79], dichloromethane [80], and the ammoniacal ethanolic solution of lipids with dimethyl ether and light petroleum in the Röse–Gottlieb extraction [81,82]. The total lipid content in samples can be determined with a gravimetric assay, Gerber-van Gulik butyrometer, infrared spectroscopy according to an International Dairy Federation (IDF) method [81], or gas chromatography equipped with a flame ionization detector [83].

To determine the MPLs and their subclasses, solid-phase extraction can fractionate polar lipids from non-polar lipids. Silica-gel-bonded cartridges or silica gel plates can be used for such a purpose [84]. The obtained MPLs can be solvent dried using a vacuum and stored at −20 °C before using [85]. In addition, chloroform and methanol are also valid elution solvents [86]. Total MPLs can be measured using the IDF molybdate assay [87], Fourier transform infrared spectroscopy [88], or a fluorescence assay on cleaved choline group [89]. Both nuclear magnetic resonance of ^{31}P and chromatography can quantify MPLs and their subclasses [90,91]. High-performance liquid chromatography coupling with detectors as a charged aerosol detector, evaporative light-scattering detector, and mass spectroscopy is more acceptable than thin layer chromatography [92].

3. Processes for Industrial Manufacturing of Milk Phospholipids

3.1. Solvent Extraction

Many kinds of polar solvents have been used to extract MPLs, such as ethanol and alkanes [21,66]. To separate casein from MPLs, proteins can also be thermally denatured or in an acid solution (pH 4.6) [48,81], the aggregated particles are subsequently filtrated. Regarding fractionation of MPL from WPPC, ethanol (70% *v/v*) at 60–80 °C denatures proteins; after filtration the MPL concentration is ≈45.8% in the filtrate in Figure 2a [48]. This notable method uses no toxic solvent. However, the incompleteness of the phospholipid recovery may be a concern [48].

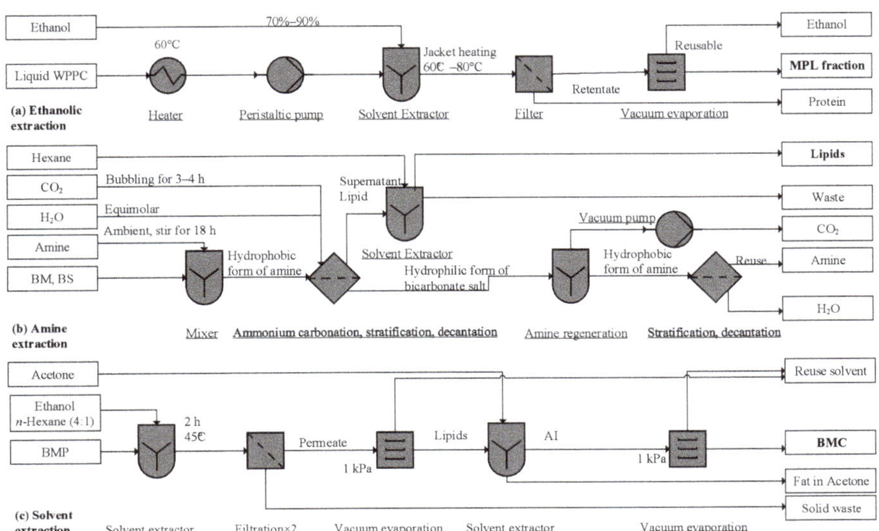

Figure 2. Process flow diagram of solvent extraction unit: (**a**) adapted from Price et al. [48], (**b**) Ota et al. [93], and (**c**) Shulman et al. [21]. BMC, buttermilk concentrate; MPL, milk phospholipid; AI, acetone insoluble; WPPC, whey protein phospholipid concentrate (liquid, reconstituted from powder).

Compared to 58.1% recovery by ethanol, the tertiary amine CyNMe2 (*N*,*N*-dimethylcycloexylamine) yielded a 99.96% recovery rate of MPLs. At various solvent–sample weight ratios, the lipid extraction was conducted at ambient temperature. The dissolved MPLs in the amine were released by bubbling

CO_2 at atmospheric pressure, which converts CyNMe2 into the carbonate salt in Figure 2b. By injecting nitrogen and removing CO_2, the carbonate salt regenerated into the amine form for reuse (Figure 2b). Though the recovery rate for BM was as high as 99.96 ± 1.2%, the extraction rates for BS and concentrated BM were only 7.57 ± 0.59% and 77.27 ± 4.51%, respectively. Aside from the input sensitivity, the amine may interact with dairy components and cause toxic consequences [93], and the chemical facilities required may be incompatible in a dairy factory setting.

MPLs can be dissolved in ethanol and alkanes [21,67,68], and may not dissolve in acetone, ethyl acetate, and 2-pentanone [21,67,68]. Lipid BMP (100 g) dissolved in ethanolic hexane (1:4 *v/v*, 800 mL) under constant agitation at 45 °C for 2 h will produce an extract. The permeate of vacuum filtration (repeated twice) can then be vacuum-dried at 1 kPa (Figure 2c). The residue (≈20 g) is then defatted twice with 120 mL acetone, and the resulting acetone is insoluble (AI, ≈7 g), composed of mainly polar lipids, and in the final step vacuum, is dried again at 1 kPa [21]. However, acetone poses a degree of toxicity, as acetone residue in defatted MPLs may reach 5–10 ppm. Further, acetone can form a mesityl oxide via a condensation reaction, causing an off flavor [94]. Hence, toxic residues in acetone-insoluble fractions need analysis and monitoring.

3.2. Supercritical Fluid Extraction

Supercritical CO_2 with ethanol as a co-solvent can be used to extract MPLs effectively, yielding purities of 26.26% and 16.88% from WPPC and BMP extractions, respectively (Figure 3a). The SFE operation can be conducted at 50–60 °C [95] and 350–550 bar [49]. The SFE co-solvent (CO_2 and 20% ethanol) allowed for complete extraction of PE, PC, and SM. However, neither PS (i.e., the vital compounds for cognitive function) nor PI were extracted [61,96]. Therefore, the co-solvent method may be an invalid industrial process due to the incompleteness of PS/PI recovery. In addition to co-solvents, dimethyl ether near the critical point (DME, 20%–30% solubility, 333 K, 40 bar) and supercritical CO_2 are able to dissolve polar and neutral lipids, respectively [59].

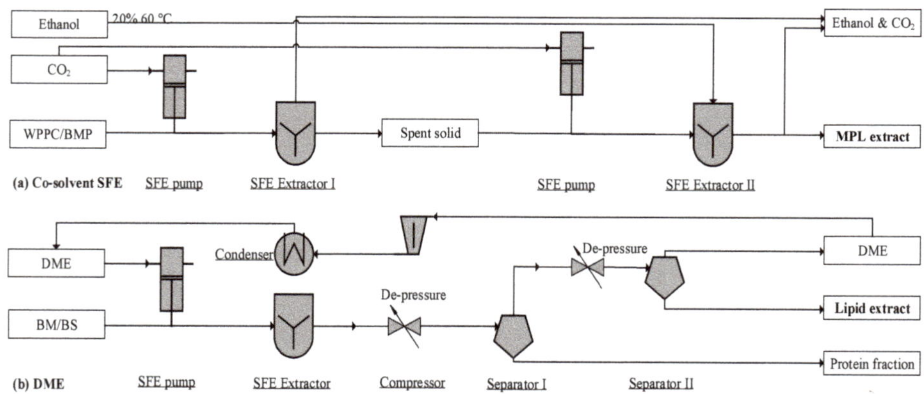

Figure 3. Process flow diagram of a supercritical fluid extraction (SFE) unit: (**a**) adapted from Li [49] and (**b**) Kala et al. [97]; WPPC, whey protein phospholipid concentrate; BS, beta serum; BM, buttermilk; BMP, buttermilk powder; DME, dimethyl ether.

Supercritical fluid DME has been used to extract polar lipids, resulting in a yield of 69.1–77.8%. The SFE process shown in Figure 3b can accept both liquid and powder inputs [59,97]. This unit can work with CO_2 and DME in two-stages, extracting neutral and polar lipids, separately. In addition to a two-step operation, this unit can also operate a single extraction with DME. Near-critical DME dissolves both polar and non-polar lipids in the SFE chamber. Through a two-stage de-pressurization, lipids are separated from the protein fraction, whereas vaporized DME is compressed and condensed for reuse (Figure 3b). This method features properties such as non-toxicity, a compact skid, feeding flexibility, and a high content of MPLs (65.7–75.5 g MPLs per 100 g DM). However, the MPL recovery rate (69.1–77.8%) needs further improvement.

3.3. Enrichment of Milk Phospholipids via Filtration

BM or BS is composed of milk fat, casein and whey protein, lactose, and ash. The particle sizes range from 0.4–4 µm for MFGM fragments or phospholipid micelles [98], 0.02–0.3 µm for casein, 0.03–0.06 µm for whey protein, and 0.002 µm for lactose and ash, respectively [99]. The size of MFG is around 0.2–15 µm [3]. As illustrated in Figure 4, the MF unit removes lactose and whey protein, and UF separates the smaller casein proteins from MPLs. Due to the size overlap of casein micelles and MPL particles, their separation is usually incomplete. Casein micelles disintegrate into peptides and amino acids in the proteolysis unit [34,42], and hydrolysates enter into the permeate stream during the subsequent UF operation [42,96]. Alcalase (E.E. 3.4.21.62), a serine-type endoprotease with esterase activity, catalyzes amino esters at pH 7.5 and 35–60 °C [96], while tryptic and peptic hydrolysis may be carried out at 42 °C for 2–16 h, at a pH of 7.7 and 2.0, respectively [42].

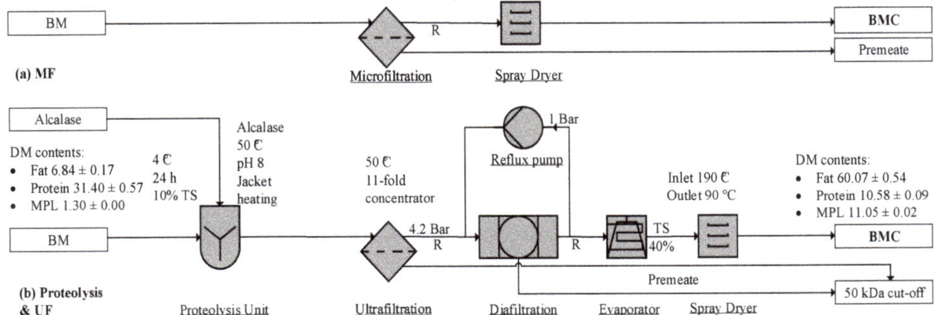

Figure 4. Filtration to enrich milk phospholipids (MPLs): (**a**) microfiltration (MF) [98] and (**b**) ultrafiltration (UF) [96]. BM, buttermilk; BMC, BM concentrate; TS, total solid; R, retentate.

Membrane filtration is a typical process for enriching BM (Figure 4a). Proteolytic treatment plus UF, as illustrated in Figure 4b, successfully differentiates MFGM from protein particles and yields product purities of 14 ± 3.4% (DM) [42] and 11.05 ± 0.02% (DM) [96]. The combined process of proteolysis and membrane separation can yield a 100% recovery rate of MPLs from BM, as illustrated in Table 3. Considering membrane units exist in most dairy factories [99,100], the process remains the most effective method for concentrating MPLs, requiring less investment than the other processes [101]. As illustrated in Table 3, this method [96] recovered more MPLs than the other processes.

Table 3. Process to purify MPLs and achieved purity (g MPLs/100 g dry product).

Reference	Input	Technology Used	Purity	Recovery (%)
[97]	BSP	SFE: CO_2, 300 bar, 40 °C, DME	12.9 → 75.7 (5.9-fold)	69.1
[97]	BSP	SFE: DME, 40 bar, 50 °C	12.9 → 66.8 (5.2-fold)	62.9
[49]	WPPC	SFE: 350 bar, CO_2, 20% ethanol, 60 °C	2.2 → 26.3 (11.9-fold)	PS/PI lost
[49]	BMP	SFE: 550 bar, CO_2, 15% ethanol, 60 °C	2.0 → 16.9 (8.6-fold)	PS/PI lost
[50]	BMC	SFE: CO_2 defat	2.2 →7.8 (3.5-fold)	100
[50]	BMC	SFE: CO_2 defat	2.2 → 9.2 (4.2-fold)	100
[98]	BMC	SFE: CO_2 defat	9.6 → 19.7 (2.1-fold)	100
[38]	BMC	SFE: CO_2 defat	7.2 → 12.0 (1.7-fold)	100
[93]	BM	Solvent: BM (6:1 *v/v*) extraction	-	87.5
[93]	BM	Solvent: BM (12:1 *v/v*) extraction	-	99.9
[93]	BS	Solvent: BS (12:1 *v/v*) extraction	-	7.6
[42]	Whey BM	Proteolysis, UF/DF, 300 kDa, 40 °C	0.3 → 8.6 (28.7-fold)	95–99
[42]	Whey BM	Proteolysis, UF/DF, 300 kDa, 40 °C	0.4 → 11.4 (27.1-fold)	95–99
[42]	Whey BM	Proteolysis, UF/DF, 300 kDa, 40 °C	0.5 → 14.0 (26.4-fold)	95–99
[96]	BMP	Proteolysis, UF/DF, 50 kDa, 50 °C	1.3 → 11.1 (8.5-fold)	100
[34]	BMP	Proteolysis, UF/DF, 50 kDa, 50 °C	0.8 → 6.2 (7.8-fold)	100
[102]	BM	MF, 0.2 μm	1.5	67
[98]	BM	MF, 0.8 μm	9.6	-
[32]	BM	MF/DF, 0.5 μm, 50 °C	1.4 → 2.5 (1.8-fold)	88.8
[32]	BM	MF/DF, 0.5 μm, 50 °C	1.4 → 4.1 (2.9-fold)	89.7
[50]	BMP	MF/DF, 0.45 μm, 9 °C	1.2 → 2.2 (1.8-fold)	60.87
[50]	BMP	MF/DF, 0.45 μm, 9 °C	1.5 → 2.2 (1.5-fold)	87.34
[50]	BMP	MF/DF, 0.45 μm, 9 °C	0.5 → 0.9 (1.7-fold)	90.12
[50]	BMP	MF/DF, 0.45 μm, 9 °C	0.3 → 0.7 (2.3-fold)	80.24
[35]	CWBM	UF, 0.15 μm cellulose acetate	1.8 → 2.3 (1.3-fold)	41.9
[35]	CWBM	UF, 0.15 μm cellulose acetate, TA	1.8 → 4.7 (2.7-fold)	98.4
[38]	CWBM	UF/DF, 10 kDa	2.0 → 7.2 (3.6-fold)	-
[36]	CWBM	TA, wash at pH 7.25, UF/DF, 55 °C	2.0 → 10.7 (5.4-fold)	>90

BM, buttermilk; BS, beta serum; BMP, BM powder; BSP, BS powder; CWBM, cheese whey BM; WPPC, whey protein phospholipid concentrate; BMC, BM concentrate; DME, dimethyl ether; SFE, supercritical fluid extraction; MF/UF/DF, micro/ultra/dia-filtration; TA, thermal aggregation.

3.4. Available Processes for Extracting Milk Phospholipids

In brief, there are three options for the large-scale manufacturing of MPLs, including solvent extraction [21,68], SFE [59,97], and proteolysis plus membrane concentration [34,42,82,96]. The membrane concentration of MPLs have yielded a 20% (*w/w*, DM basis) purity, as achieved by Lecico [58] and Arla [10]. Tatua [56] and Westland and Synlait [44] have extracted MPLs from BS powder (2.28%, *w/w*, DM basis), achieving approximately 12.8% (*w/w*, DM basis) purity using membrane filtration. The proteolysis and UF unit recovers MPLs completely [34,82,96] and cost-effectively [44]. This process is more efficient than SFE and solvent extraction, whereas SFE and solvent extractions are effective steps for manufacturing high purity MPLs. Therefore, the three processes have features of a high recovery rate, facility availability, and food compatibility, representing current industrial practices (in Table 3).

4. Carbon Footprint

4.1. Life-Cycle Accessment Method of Carbon Footprint

The ISO 14,040 life-cycle assessment (LCA) is an internationally accepted methodology used to calculate a product's environmental footprint [103]. Life-cycle carbon footprints (CFs) of dairy products cover the direct emission from the dairy factory (scope 1); the energy carrier footprint for factory operations (natural gas, steam, power, nitrogen, and compressed air in scope 2); and the raw material, packaging, and logistics in scope 3. In addition, the life-cycle CF comprises the emissions from the dairy farm (upstream) and distribution center (downstream) [104]. The boundaries are set as shown in Figure 5a. The CFs of MPL products were reported as equivalent CO_2 emission for one kg of MPLs, according to the ISO 14,067 reporting standards [105].

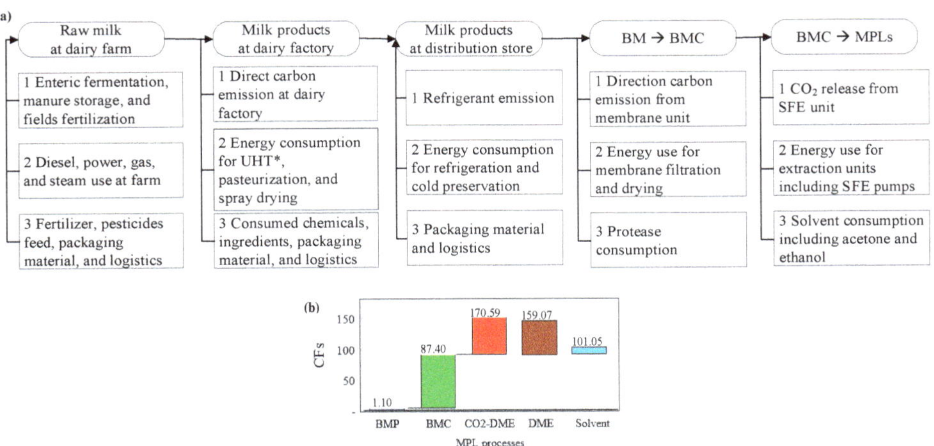

Figure 5. (**a**) Boundary definition of the life-cycle carbon footprints (CFs) of dairy products and exemplary emissions from scope 1 (direction emission), scope 2 (energy carriers), and scope 3 (raw material procured, packaging material, and transportation). (**b**) Cascade of CFs of BMP, BMC, and MPLs using the following processes: "CO_2-DEM" (supercritical CO_2 and DME), "DME" (supercritical DME), and "Solvent" (hexane and ethanol extraction and acetone de-fatting, kg equivalent CO_2/products). Scopes of BM CFs: adapted from References [45,106–109]. MPLs, milk phospholipids; BM, buttermilk; BMC, BM concentrate; DME, dimethyl ether; SFE, supercritical fluid extraction. * Ultra High Temperature processing.

The CF of BM (baseline CF, 1.10 kg CO_2/kg BM powder) was cited directly from data derived from the Unified Livestock Industry and Crop Emissions Estimation System (ULICEES) model in Canada [45]. The data abides by the Intergovernmental Panel on Climate Change (IPCC) methodology [106]; it covers emissions like methane [45], nitrous oxide [107], and carbon dioxide using the F4E2 model [108]; and uses an allocation matrix to partition six inventory flows (i.e., fuel, power, raw milk transportation, alkaline/acid, water, and waste water) into 11 major dairy products [109].

In this study, BM was assumed as the starting material for producing MPLs. Therefore, the CF for producing BM was set as the baseline. The CF of MPLs in Figure 5b and Table 4 is a sum of the CF of BM (as the baseline) and the CF for extracting MPLs from BM at dairy factories. The starting amount of BM was assumed to be 100 kg (1.3%, *w/w*, DM basis). Since MPLs are considered as the target products, CF of protein in the MPL fractions was not included in the estimations.

The CF of BM concentrate (BMC) via membrane separation (MS) was calculated using the equation: $CF_{BMC} = CF_{BM} + CF_{MS}$, where CF_{BMC}, CF_{BM}, and CF_{MS} were the CF of BMC, BM, and MS, respectively. The CF of MPL products using SFE or solvent extraction was calculated using the equation $CF_{MPLs} = CF_{BMC} + CF_{SFE}$ or $CF_{MPLs} = CF_{BMC} + CF_{Sol}$ where CF_{MPLs}, CF_{BMC}, CF_{SFE}, and CF_{Sol} were the CF for the MPL product, BMC, SFE, and solvent extraction process, respectively, as illustrated in Figure 5b. The CF for BMP (1.5% purity) was 1.10 kg CO_2/kg MPL, and the CF of BMC (11.0% purity) was 87.40 kg CO_2/kg MPL, as calculated in Table 5. Starting from BMC, the CFs of MPL products were 170.59, 159.07, and 101.05 kg CO_2/kg MPL for processes of CO_2-DME supercritical fluid extraction, DME SFE, and organic solvent extraction, respectively.

Table 4. Normalized carbon footprints of milk phospholipids (kg CO_2/kg MPLs).

Process	Membrane	SFE (CO_2/DME)	SFE (DME)	Solvent Extract	Unit
Reference	[96]	[59,97]	[59,97]	[21]	-
Input	BMP	BMC	BMC	BMC	-
Input amount	100.00	100.00	100.00	100.00	kg
Input purity	1.3	5.7	6.8	12.3	g/100 g DM
Product	BMC	MPLs	MPLs	MPLs	-
Product amount	11.76	5.13	6.56	13.98	kg
Product purity	11.05	76.80	66.80	88.00	g/100 g DM
MPL yield	100.00	69.10	67.40	100.00	%
Power	17.48	512.85	655.68	-	kWh
Material used	Alcalase 0.03	CO_2 1000.00 DME 200.00	DME 200.00	C_6/ethanol 552.00 Acetone 189.60	kg kg
Thermal energy	13.10	-	-	-	MJ
Power CF factor	0.1567	0.1567	0.1567	0.1567	kg CO_2/kWh
Material CF factor	5.00	CO_2 0.05 DME 0.16	CO_2 0.05 DME 0.16	C_6/ethanol 0.16 Acetone 0.42	kg CO_2/kg kg CO_2/kg
Thermal CF factor	0.06	-	-	-	kg CO_2/MJ
CF of power	2.74	80.36	102.74	-	kg CO_2
CF of material	0.16	82.00	32.00	167.95	kg CO_2
Thermal CF	0.72	-	-	-	kg CO_2
Utility CF	3.62	162.36	134.74	167.95	kg CO_2
BM/BMC baseline	110.00	498.17	594.31	1074.99	kg CO_2
Product CF	9.66	128.80	111.19	88.93	kg CO_2/kg
Normalized CF	87.40	170.59	159.07	101.05	kg CO_2/kg MPLs

BM, buttermilk; BMC, BM, concentrate; MPLs, milk phospholipids; C6, hexane; DME, dimethyl ether; SFE, supercritical fluid extraction; UF/DF, ultra/dia-filtration; CF, carbon footprint. Membrane filtration power consumption: 1.486 kWh/kg products [110]; Canada power CF factor: 0.1567 kg CO_2/kWh [111]; CF of reusable solvents (DME, hexane and ethanol): 0.16 kg CO_2/kg solvent; reused acetone CF: 0.42 kg CO_2/kg solvent [112]; DME CF 1.01 kg/kg; 84% reuse [113]; baseline of BM: 1.1 kg CO_2/kg BMP [45]; SFE CO_2 reuse 95% [114]; SFE CO_2/DME power cost estimation 100 kWh/kg extract [115].

4.2. Carbon Footprint Estimation

In Table 4, four MPL enrichment processes were used as references for estimating and comparing the total CFs. The membrane separation process was used to concentrate MPL from the original BM. The resulting product was a BM concentrate (BMC), which may be further processed to yield MPL products by either using an SFE technique or a solvent extraction method. The CF of "utility" consumed for the three individual MPL enrichment methods was obtained by multiplying the utility amount and CF conversion factor, which represents the amount of carbon emission for a unit weight of utility. Normalized CF: $CF_{Normalized} = CF/C_{MPLs}$, where CMPLs was the MPL purity (g MPLs per 100 g product).

The normalized CF of the product uisng membrane separation was as high as 87.4 kg CO_2/kg BMC since the BMC comprised of only 11.05% MPLs. The CFs for products using SFE and solvent extraction were much higher than their baseline (CF_{BMC}) because of the intensive process during purification. As shown in Table 4, the CFs of fractions using SFE were 170.59 and 159.07 kg CO_2/kg MPLs for CO_2/DME co-extraction and DME extraction, respectively. CO_2/DME co-SFE exhibited a higher environmental impact compared to supercritical DME extraction due to direct emissions from co-SFE. Solvent extraction demonstrated a lower environmental impact and a higher MPL recovery rate than SFE. However, the products obtained using solvent extraction were less food-compatible than SFE unit-extracted products.

MPLs from proteolysis and filtration processes carry 87.40 kg equivalent CO_2/kg product, much higher than all the milk fat products (Table 5). With less CF than SFE and solvent extraction, membrane separation is the most efficient process in terms of process intensity, energy consumption, and environmental impact. In addition, this process is compatible with most dairy factories. Membrane separation is a necessary step for concentrating BM into BMC. BMC can then be purified using SFE (DME). The relevant processes with a significant MPL CF include membrane filtration, evaporation and spray drying, SFE, and solvent recovery, the improvement of which offer opportunities to reduce the CF of the final products. For example, 0.1-μm polymeric spiral-wound MF membranes have been used to separate casein from milk, exhibiting a higher energy efficiency at 0.024 (MF) and 0.015 (DF) kWh/kg permeate than that of graded permeability membrane (0.143 and 0.077 kWh/kg permeate for MF and DF, respectively [110]. Furthermore, permeate flux, volume concentration ratio, transmembrane pressure, and temperature all had an impact on the energy efficiency of membrane UF, ranging from 0.26–0.33 kWh/kg retentate [116]. Another approach toward reducing the environmental impact is to improve the purity of MPLs during filtration by differentiating the particle size of casein micelles (i.e., hydrolysis) from the fragmented MFGM and subsequent application of membrane filtration.

Table 5. Comparison of the carbon footprint of milk phospholipids in commercial dairy products (kg CO_2/kg product).

Dairy Products	CF	Scope 1	Scope 2	Scope 3	Country	Reference
Raw milk	1.10	-	-	-	Canada	[45]
Bulk liquid	1.00	0.870	0.065	0.065	Canada	[45]
Yogurt	1.50	1.083	0.252	0.165	Canada	[45]
Whole milk	1.12	0.843	0.173	0.104	China	[117]
Powder milk	10.10	-	-	-	Canada	[45]
Butter	7.30	-	-	-	Canada	[45]
BM	1.10	-	-	-	Canada	[45]
Cheese	12.40	-	-	-	Italy	[104]
Cheese	5.30	-	-	-	Canada	[45]
Cheese	8.80	-	-	-	Sweden	[118]
BM → BMC: UF/DF	87.40	-	-	-	-	[96]
BM → BMC → MPLs: SFE CO_2/DME	170.59	-	-	-	-	[97]
BM → BMC → MPLs: SFE DME	159.07	-	-	-	-	[97]
BM → BMC → MPLs: Solvent extract	101.05	-	-	-	-	[45]

MPLs, milk phospholipids; BM, buttermilk; BMC, BM concentrate; DME, dimethyl ether; SFE, supercritical fluid extraction; UF/DF, ultra/dia-filtration; CFs, carbon footprints.

5. Conclusions

This paper identified three dairy streams for milk phospholipid (MPL) manufacturing at an industrial scale: buttermilk, beta serum, and whey protein phospholipid concentrate. The life-cycle CFs of the MPLs were 87.40, 170.59, 159.07, and 101.05 kg CO_2/kg MPLs for the membrane separation process, CO_2/DME supercritical fluid extraction, SFE by DME, and organic solvent extraction, respectively. The extracted products comprised 11.1, 76.8, 69.9, and 88.0% MPLs, with recovery rate of 100, 69.1, 67.4, and 100%, respectively. In conclusion, to improve the efficiency of an MPL concentration process, casein in BM needs to be proteolyzed before running UF/DF processes. By doing so, it is possible to achieve full recovery of MPLs from BM; moreover, this method may result in a relatively low CF. SFE using dimethyl ether is the most effective method for the production of high-purity (≈66.8%) MPL products, albeit at the cost of a high CF. This study provided insights into the best available industrial practices for extracting MPLs and estimating their life-cycle CFs.

Author Contributions: Conceptualization, Z.H., H.Z., and C.S.B.; investigation, Z.H.; writing—original draft preparation, Z.H.; writing—review and editing, L.S., H.Z., M.S.M., C.S.B., L.L., and D.K. All authors have read and agreed to the published version of the manuscript.

Funding: This research received no external funding.

Conflicts of Interest: The authors declare no conflicts of interest.

References

1. Ortega-Anaya, J.; Jiménez-Flores, R. Symposium review: The relevance of bovine milk phospholipids in human nutrition—Evidence of the effect on infant gut and brain development. *J. Dairy Sci.* **2018**, *102*, 1–11. [CrossRef]
2. Verardo, V.; Arráez-Román, A.M.G.-C.D.; Hettinga, K. Recent advances in phospholipids from colostrum, milk and dairy by-products. *Int. J. Mol. Sci.* **2017**, *18*, 173. [CrossRef] [PubMed]
3. Zheng, H.; Jimenez-Flores, R.; Gragson, D.; Everett, D.W. Phospholipid architecture of the bovine milk fat globule membrane using giant unilamellar vesicles as a model. *J. Agric. Food Chem.* **2014**, *62*, 3236–3243. [CrossRef] [PubMed]
4. Zheng, H.; Jiménez-Flores, R.; Everett, D.W. Lateral lipid organization of the bovine milk fat globule membrane is revealed by washing processes. *J. Dairy Sci.* **2014**, *97*, 5964–5974. [CrossRef] [PubMed]
5. Holzmüller, W.; Kulozik, U. Technical difficulties and future challenges in isolating membrane material from milk fat globules in industrial settings—A critical review. *Int. Dairy J.* **2016**, *61*, 51–66. [CrossRef]
6. Lopez, C.; Blot, M.; Briard-Bion, V.; Cirie, C.; Graulet, B. Butter serums and buttermilks as sources of bioactive lipids from the milk fat globule membrane: Differences in their lipid composition and potentialities of cow diet to increase n-3 PUFA. *Food Res. Int.* **2017**, *100*, 864–872. [CrossRef] [PubMed]
7. Liu, Z.; Logan, A.; Cocks, B.G.; Rochfort, S. Seasonal variation of polar lipid content in bovine milk. *Food Chem.* **2017**, *237*, 865–869. [CrossRef]
8. Gassi, J.Y.; Blot, M.; Beaucher, E.; Robert, B.; Leconte, N.; Camier, B.; Rousseau, F.; Bourlieu, C.; Jardin, J.; Briard-Bion, V.; et al. Preparation and characterisation of a milk polar lipids enriched ingredient from fresh industrial liquid butter serum: Combination of physico-chemical modifications and technological treatments. *Int. Dairy J.* **2016**, *52*, 26–34. [CrossRef]
9. Burling, H.; Andersson, I.; Schneider, M. Phosphatidylserine Enriched Milk Fractions for the Formulation of Functional Foods. U.S. Patent Application No. 8231922B2, 31 July 2012.
10. Burling, H.; Graverholt, G. Milk—A new source for bioactive phospholipids for use in food formulations. *Lipid Technol.* **2008**, *20*, 229–231. [CrossRef]
11. Castro-Gómez, P.; Garcia-Serrano, A.; Visioli, F.; Fontecha, J. Relevance of dietary glycerophospholipids and sphingolipids to human health. *Prostaglandins Leukot. Essent. Fat. Acids* **2015**, *101*, 41–51. [CrossRef]
12. Burling, H.; Andersson, I.; Schneider, M. Phosphatidylserine Enriched Milk Fractions for the Formulation of Functional Foods. W.O. Patent Application No. 128465A1, 7 December 2006.
13. Kuchta-Noctor, A.M.; Murray, B.A.; Stanton, C.; Devery, R.; Kelly, P.M. Anticancer activity of buttermilk against SW480 colon cancer cells is associated with caspase-independent cell death and attenuation of Wnt, Akt, and ERK signaling. *Nutr. Cancer* **2016**, *68*, 1234–1246. [CrossRef] [PubMed]
14. Contarini, G.; Povolo, M. Phospholipids in milk fat: Composition, biological and technological significance, and analytical strategies. *Int. J. Mol. Sci.* **2013**, *14*, 2808–2831. [CrossRef] [PubMed]
15. Hernell, O.; Timby, N.; Domellöf, M.; Lönnerdal, B. Clinical benefits of milk fat globule membranes for infants and children. *J. Pediatr.* **2016**, *173*, S60–S65. [CrossRef] [PubMed]
16. Timby, N.; Hernell, O.; Vaarala, O.; Melin, M.; Lönnerdal, B.; Domellöf, M. Infections in infants fed formula supplemented with bovine milk fat globule membranes. *J. Pediatr. Gastroenterol. Nutr.* **2015**, *60*, 384–389. [CrossRef] [PubMed]
17. Küllenberg, D.; Taylor, L.A.; Schneider, M.; Massing, U. Health effects of dietary phospholipids. *Lipids Health Dis.* **2012**, *11*, 1–16. [CrossRef] [PubMed]
18. Rutenberg, D. Infant Formula Supplemented with Phospholipids. W.O. Patent Application No. 105609A1, 24 December 2003.
19. Timby, N.; Domellöf, M.; Lönnerdal, B.; Hernell, O. Supplementation of infant formula with bovine milk fat globule membrane. *Adv. Nutr.* **2017**, *8*, 351–355. [CrossRef]
20. Braak, V.D.; Maria, C.C.; Thomassen, G.; Acton, D.S.; Abrahamse, E. Nutrition with Large Lipid Globules Comprising Vegetable Fat Coated with Milk Phospholipids for Lipid Digestion. W.O. Patent Application No. 163881A1, 13 Octorber 2016.
21. Shulman, A.; Zuabi, R.; Dror, G.B.; Twito, Y.; Pelled, D.; Herzog, Y. Polar Lipid Mixtures, Their Preparation and Uses. U.S. Patent Application No. 9814252B2, 8 August 2017.

22. Rueda, R.; Barranco, A.; Ramirez, M.; Vazquez, E.; Valverde, E.; Prieto, P.; Dohnalek, M.H. Enriched Infant Formulas. U.S. Patent Application No. 0057178A1, 6 March 2008.

23. Lopez, C.; Menard, O. Human milk fat globules: Polar lipid composition and in situ structural investigations revealing the heterogeneous distribution of proteins and the lateral segregation of sphingomyelin in the biological membrane. *Colloids Surf. B Biointerfaces* **2011**, *83*, 29–41. [CrossRef]

24. Thompson, A.K.; Haisman, D.; Singh, H. Physical stability of liposomes prepared from milk fat globule membrane and soya phospholipids. *J. Agric. Food Chem.* **2006**, *5*, 6390–6397. [CrossRef]

25. Huang, Z.; Stipkovits, L.; Zheng, H.; Serventi, L.; Brennan, C.S. Bovine milk fats and their replacers in baked goods: A review. *Foods* **2019**, *8*, 383. [CrossRef]

26. Vanderghem, C.; Deroanne, C.; Bodson, P.; Blecker, C. Milk fat globule membrane and buttermilks: From composition to valorization. *Biotechnol. Agron. Soc. Environ.* **2010**, *14*, 485–500.

27. Phan, T.T.Q.; Le, T.T.; Van de Walle, D.; Van der Meeren, P.; Dewettinck, K. Combined effects of milk fat globule membrane polar lipids and protein concentrate on the stability of oil-in-water emulsions. *Int. Dairy J.* **2016**, *52*, 42–49. [CrossRef]

28. Huppertz, T.; Kelly, A. Physical chemistry of milk fat globules. In *Advanced Dairy Chemistry Volume 2 Lipids*; Springer: New York, NY, USA, 2006; pp. 173–212.

29. Britten, M.; Lamothe, S.; Robitaille, G. Effect of cream treatment on phospholipids and protein recovery in butter-making process. *Int. J. Food Sci. Technol.* **2008**, *43*, 651–657. [CrossRef]

30. Rombaut, R.; Camp, J.V.; Dewettinck, K. Phospho- and sphingolipid distribution during processing of milk, butter and whey. *Int. J. Food Sci. Technol.* **2006**, *41*, 435–443. [CrossRef]

31. Lamothe, S.; Robitaille, G.; St-Gelais, D.; Britten, M. Butter making from caprine creams: Effect of washing treatment on phospholipids and milk fat globule membrane proteins distribution. *J. Dairy Res.* **2008**, *75*, 439–443. [CrossRef]

32. Morin, P.; Britten, M.; Jimenez-Flores, R.; Pouliot, Y. Microfiltration of buttermilk and washed cream buttermilk for concentration of milk fat globule membrane components. *J. Dairy Sci.* **2007**, *90*, 2132–2140. [CrossRef]

33. Zheng, H.; Jiménez-Flores, R.; Everett, D.W. Bovine milk fat globule membrane proteins are affected by centrifugal washing processes. *J. Agric. Food Chem.* **2013**, *61*, 8403–8411. [CrossRef]

34. Barry, K.M.; Dinan, T.G.; Kelly, P.M. Selective enrichment of dairy phospholipids in a buttermilk substrate through investigation of enzymatic hydrolysis of milk proteins in conjunction with ultrafiltration. *Int. Dairy J.* **2017**, *68*, 80–87. [CrossRef]

35. Rombaut, R.; Dejonckheere, V.; Dewettinck, K. Filtration of milk fat globule membrane fragments from acid buttermilk cheese whey. *J. Dairy Sci.* **2007**, *90*, 1662–1673. [CrossRef]

36. Rombaut, R.; Dewettinck, K. Thermocalcic aggregation of milk fat globule membrane fragments from acid buttermilk cheese whey. *J. Dairy Sci.* **2007**, *90*, 2665–2674. [CrossRef]

37. Levin, M.A.; Burrington, K.J.; Hartel, R.W. Composition and functionality of whey protein phospholipid concentrate and delactosed permeate. *J. Dairy Sci.* **2016**, *99*, 6937–6947. [CrossRef]

38. Costa, M.R.; Elias-Argote, X.E.; Jiménez-Flores, R.; Gigante, M.L. Use of ultrafiltration and supercritical fluid extraction to obtain a whey buttermilk powder enriched in milk fat globule membrane phospholipids. *Int. Dairy J.* **2010**, *20*, 598–602. [CrossRef]

39. Food and Agriculture Organization-World Health Organization. *Milk and Milk Products*; World Health Organization, Food and Agriculture Organization of the United Nations: Rome, Italy, 2007.

40. Dairymark. *Buttermilk—A Strategic Review of Opportunities and Applications*; Shainwright Consulting and Research Group Pty Ltd.: Norwood, Australia, 2007; p. 84.

41. United States Department of Agriculture. *National Nutrient Database for Standard Reference Software v.3.9.5.3*; United States Department of Agriculture, Ed.; United States Department of Agriculture: Washington, DC, USA, 1 April 2018. Available online: http://ndb.nal.usda.gov/ndb (accessed on 6 January 2020).

42. Konrad, G.; Kleinschmidt, T.; Lorenz, C. Ultrafiltration of whey buttermilk to obtain a phospholipid concentrate. *Int. Dairy J.* **2013**, *30*, 39–44. [CrossRef]

43. Lambert, S.; Leconte, N.; Blot, M.; Rousseau, F.; Robert, B.; Camier, B.; Gassi, J.-Y.; Cauty, C.; Lopez, C.; Gésan-Guiziou, G. The lipid content and microstructure of industrial whole buttermilk and butter serum affect the efficiency of skimming. *Food Res. Int.* **2016**, *83*, 121–130. [CrossRef]

44. Ireland, E.R. Developing a Better Buttermilk Solution. Master's Thesis, Engineering Management, University of Canterbury, Canterbury, New Zealand, 2014.
45. Verge, X.P.; Maxime, D.; Dyer, J.A.; Desjardins, R.L.; Arcand, Y.; Vanderzaag, A. Carbon footprint of Canadian dairy products: Calculations and issues. *J. Dairy Sci.* **2013**, *96*, 6091–6104. [CrossRef]
46. Ali, A.H. Current knowledge of buttermilk: Composition, applications in the food industry, nutritional and beneficial health characteristics. *Int. J. Dairy Technol.* **2018**, *72*, 169–182. [CrossRef]
47. Yildiz, N.; Bakirci, I. Investigation of the use of whey powder and buttermilk powder instead of skim milk powder in yogurt production. *J. Food Sci. Technol.* **2019**, *56*, 4429–4436. [CrossRef]
48. Price, N.; Fei, T.; Clark, S.; Wang, T. Extraction of phospholipids from a dairy by-product (whey protein phospholipid concentrate) using ethanol. *J. Dairy Sci.* **2018**, *101*, 8778–8787. [CrossRef]
49. Li, B. Selective Extraction of Phospholipids from Dairy Powders Using Supercritical Fluid Extraction. Ph.D. Thesis, Kansas State University, Manhattan, KS, USA, 2017.
50. Spence, A.J.; Jimenez-Flores, R.; Qian, M.; Goddik, L. Phospholipid enrichment in sweet and whey cream buttermilk powders using supercritical fluid extraction. *J. Dairy Sci.* **2009**, *92*, 2373–2381. [CrossRef]
51. Fauquant, J.; Beaucher, E.; Sinet, C.; Robert, B.; Lopez, C. Combination of homogenization and cross-flow microfiltration to remove microorganisms from industrial buttermilks with an efficient permeation of proteins and lipids. *Innov. Food Sci. Emerg. Technol.* **2014**, *21*, 131–141. [CrossRef]
52. Thompson, A. Structure and Properties of Liposomes Prepared from Milk Phospholipids. Ph.D. Thesis, Massey University, Palmerston North, New Zealand, 2005.
53. Li, Z. Encapsulation of Bioactive Salmon Protein Hydrolysates with Chitosan-Coated Liposomes. Master's Thesis, Master of Science, Dalhousie University, Halifax, NS, Canada, 2014.
54. Sokol, E.; Ulven, T.; Faergeman, N.J.; Ejsing, C.S. Comprehensive and quantitative profiling of lipid species in human milk, cow milk and a phospholipid-enriched milk formula by GC and MS/MS(ALL). *Eur. J. Lipid Sci. Technol.* **2015**, *117*, 751–759. [CrossRef]
55. Arla Foods Ingredients: Ingredients for the Next Generation. Available online: https://www.foodingredientsfirst.com/supplier-profiles/Arla-Foods-Ingredients.html (accessed on 27 February 2020).
56. Tatua Specialty Nutritional Ingredients: Phospholipids. Available online: https://www.tatua.com/specialty-nutritionals-ingredients/phospholipids/ (accessed on 1 January 2020).
57. Moukarzel, S. The Complexity of Understanding Human Milk Components and Infant Brain Development. Ph.D. Thesis, Human Nutrition, University of British Columbia, Vancouver, BC, Canada, 2016.
58. Lecico Milk Phospholipids. Available online: http://www.lecico.de/en/products/milk-phospholipids (accessed on 23 January 2020).
59. Fletcher, K.; Catchpole, O.; Grey, J.B.; Pritchard, M. Beta-Serum Dairy Products, Neutral Lipid-Depleted and/or Polar Lipid-Enriched Dairy Products, and Processes for Their Production. U.S. Patent Application No. 8471002B2, 25 June 2013.
60. Meggle. Composition Richly Containing Polar Lipid and Method of Manufacturing the Same. E.P. Patent Application No. 2168438A1, 31 March 2010.
61. Catchpole, O.J.; Tallon, S.J. Process for Separation Lipid Materials. W.O. Patent Application No. 123424A1, 20 April 2007.
62. Dalemans, D.; Blecker, C.; Bodson, P.; Danthine, S.; Deroanne, C.; Paquot, M. Milk Ingredient Enriched in Polar Lipids and Uses Thereof. U.S. Patent Application No. 0068293A1, 18 March 2010.
63. Dwwettinck, K.; Boone, M. Method for Obtaining Products Enriched in Phospho- and Sphingolipids. W.O. Patent Application No. 0234062A1, 2 May 2002.
64. Gnanasambandam, R.; Patel, H. Methods of Concentrating Phospholipds. U.S. Patent Application No. 0335778A1, 7 November 2018.
65. Sato, I. Method of Separation and Recovery of Lipid in Membrane Material of Fat Globule. JP Patent Application No. 336230A, 8 December 2005.
66. Suzuki, A.; Shioda, M.; Imai, M. Manufacturing Method of Sphingoid Base-Containing Extract. JP Patent Application No. 052912A, 29 August 2018.
67. Bruecher, T.; Demey, J.; Katte, M.; Molnar, J.; Tirok, S. Method for the Fractionation of Phospholipids from Phospholipid-Containing Material. U.S. Patent Application No. 9567356B2, 14 February 2017.
68. Nyberg, L.; Burling, H. Method for Extracting Sphingomyelin. U.S. Patent Application No. 5677472, 14 October 1997.

69. Dewettinck, K.; Rombaut, R.; Thienpont, N.; Le, T.T.; Messens, K.; Van Camp, J. Nutritional and technological aspects of milk fat globule membrane material. *Int. Dairy J.* **2008**, *18*, 436–457. [CrossRef]

70. Lu, J.; Argov-Argama, N.; Anggrek, J.; Boeren, S.; Hooijdonk, T.V.; Vervoort, J.; Hetting, K.A. The protein and lipid composition of the membrane of milk fat globules depends on their size. *J. Dairy Sci.* **2016**, *99*, 4726–4738. [CrossRef] [PubMed]

71. Jing, L. The Biology of Milk Synthesis from a Proteomics Perspective. Ph.D. Thesis, Wageningen University, Wageningen, Holland, 2013.

72. Bezelgues, J.B.; Morgan, F.; Palomo, G.; Crosset-Perrotin, L.; Ducret, P. Short communication: Milk fat globule membrane as a potential delivery system for liposoluble nutrients. *J. Dairy Sci.* **2009**, *92*, 2524–2528. [CrossRef] [PubMed]

73. Holzmüller, W.; Müller, M.; Himbert, D.; Kulozik, U. Impact of cream washing on fat globules and milk fat globule membrane proteins. *Int. Dairy J.* **2016**, *59*, 52–61. [CrossRef]

74. Spitsberg, V.L.; Ivanov, L.; Shritz, V. Recovery of milk fat globule membrane (MFGM) from buttermilk: Effect of Ca-binding salts. *J. Dairy Res.* **2019**, *86*, 374–376. [CrossRef]

75. Holzmüller, W.; Kulozik, U. Isolation of milk fat globule membrane (MFGM) material by coagulation and diafiltration of buttermilk. *Int. Dairy J.* **2016**, *63*, 88–91. [CrossRef]

76. Hansen, S.F.; Hogan, S.A.; Tobin, J.; Rasmussen, J.T.; Larsen, L.B.; Wiking, L. Microfiltration of raw milk for production of high-purity milk fat globule membrane material. *J. Food Eng.* **2020**, *276*, 109887. [CrossRef]

77. Bourlieua, C.; Cheillan, D.; Blota, M.; Daira, P.; Trauchessec, M.; Ruet, S.; Gassi, J.-Y.; Beaucher, E.; Robert, B.; Leconte, N.; et al. Polar lipid composition of bioactive dairy co-products buttermilk and butterserum: Emphasis on sphingolipid and ceramide isoforms. *Food Chem.* **2018**, *240*, 67–74. [CrossRef]

78. Cheema, M.; Smith, P.B.; Patterson, A.D.; Hristov, A.; Hart, F.M. The association of lipophilic phospholipids with native bovine casein micelles in skim milk: Effect of lactation stage and casein micelle size. *J. Dairy Sci.* **2017**, *101*, 8672–8687. [CrossRef]

79. Gallier, S.; Gragson, D.; Cabral, C.; Jimenez-Flores, R.; Everett, D.W. Composition and fatty acid distribution of bovine milk phospholipids from processed milk products. *J. Agric. Food Chem.* **2010**, *58*, 10503–10511. [CrossRef]

80. Claumarchirant, L.; Cilla, A.; Matencio, E.; Sanchez-Siles, L.M.; Castro-Gomez, P.; Fontecha, J.; Alegría, A.; Lagarda, M.J. Addition of milk fat globule membrane as an ingredient of infant formulas for resembling the polar lipids of human milk. *Int. Dairy J.* **2016**, *61*, 228–238. [CrossRef]

81. Ferreiro, T.; Martínez, S.; Gayoso, L.; Rodríguez-Otero, J.L. Evolution of phospholipid contents during the production of quark cheese from buttermilk. *J. Dairy Sci.* **2016**, *99*, 4154–4159. [CrossRef] [PubMed]

82. Barry, K.M.; Dinan, T.G.; Murray, B.A.; Kelly, P.M. Comparison of dairy phospholipid preparative extraction protocols in combination with analysis by high performance liquid chromatography coupled to a charged aerosol detector. *Int. Dairy J.* **2016**, *56*, 179–185. [CrossRef]

83. Rodríguez-Alcal, L.M.; Castro-Gomez, P.; Felipe, X.; Noriega, L.; Fontecha, J. Effect of processing of cowmilk by high pressures under conditions up to 900 MPa on the composition of neutral, polar lipids and fatty acids. *LWT Food Sci. Technol.* **2015**, *62*, 265–270. [CrossRef]

84. Zou, X.; Guo, Z.; Jin, Q.; Huang, J.; Cheong, L.; Xu, X.; Wang, X. Composition and microstructure of colostrum and mature bovine milk fat globule membrane. *Food Chem.* **2015**, *185*, 362–370. [CrossRef]

85. Haddadian, Z.; Eyres, G.T.; Bremer, P.; Everett, D.W. Polar lipid composition of the milk fat globule membrane in buttermilk made using various cream churning conditions or isolated from commercial samples. *Int. Dairy J.* **2018**, *81*, 138–142. [CrossRef]

86. Walczak, J.; Pomastowski, P.; Bocian, S.; Buszewski, B. Determination of phospholipids in milk using a new phosphodiester stationary phase by liquid chromatography-matrix assisted desorption ionization mass spectrometry. *J. Chromatogr. A* **2016**, *1432*, 39–48. [CrossRef]

87. Vilamarim, R.; Bernardo, J.; Videira, R.A.; Valentão, P.; Veiga, F.; Andrade, P.B. An egg yolk's phospholipid-pennyroyal nootropic nanoformulation modulates monoamino oxidase-A (MAO-A) activity in SH-SY5Y neuronal model. *J. Funct. Foods* **2018**, *46*, 335–344. [CrossRef]

88. Kala, R.; Samková, E.; Pecová, L.; Hanuš, O.; Sekmokas, K.; Riaukienė, D. An overview of determination of milk fat: Development, quality control measures, and application. *Acta Univ. Agric. Silvic. Mendel. Brun.* **2018**, *66*, 1055–1064. [CrossRef]

89. Shrestha, P.; Davis, D.A.; Veeranna, R.P.; Carey, R.F.; Viollet, C.; Yarchoan, R. Hypoxia-inducible factor-1 alpha as a therapeutic target for primary effusion lymphoma. *PLoS Pathog.* **2017**, *13*, e1006628. [CrossRef]

90. Hickey, C.D.; Diehl, B.W.K.; Nuzzo, M.; Millqvist-Feurby, A.; Wilkinson, M.G.; Sheehana, J.J. Influence of buttermilk powder or buttermilk addition on phospholipid content, chemical and bio-chemical composition and bacterial viability in Cheddar style-cheese. *Food Res. Int.* **2017**, *102*, 748–758. [CrossRef] [PubMed]

91. Xu, S.; Walkling-Ribeiro, M.; Griffiths, M.W.; Corredig, M. Pulsed electric field processing preserves the antiproliferative activity of the milk fat globule membrane on colon carcinoma cells. *J. Dairy Sci.* **2015**, *98*, 2867–2874. [CrossRef] [PubMed]

92. Rodríguez-Alcalá, L.M.; Fontecha, J. Major lipid classes separation of buttermilk, and cows, goats and ewes milk by high performance liquid chromatography with an evaporative light scattering detector focused on the phospholipid fraction. *J. Chromatogr. A* **2010**, *1217*, 3063–3066. [CrossRef] [PubMed]

93. Cheng, S.; Rathnakumar, K.; Martinez-Monteagudo, S.I. Extraction of dairy phospholipids using switchable solvents: A feasibility study. *Foods* **2019**, *8*, 265. [CrossRef]

94. Hutton, K.J.; Guymon, J.S. Process for Producing Deoiled Phosphatides. CA Patent Application No. 2354705, 20 October 2009.

95. Ota, M.; Oda, E.; Kataoka, S.; Sato, Y.; Inomata, H. Supercritical fluid extraction of high-value natural products from buttermilk analyzed by a dynamic extraction model. *Nippon Shokuhin Kagaku Kogaku Kaishi* **2018**, *65*, 251–258. [CrossRef]

96. Barry, K.M.; Dinan, T.G.; Kellya, P.M. Pilot scale production of a phospholipid-enriched dairy ingredient by means of an optimised integrated process employing enzymatic hydrolysis, ultrafiltration and super-critical fluid extraction. *Innov. Food Sci. Emerg. Technol.* **2017**, *41*, 301–306. [CrossRef]

97. Catchpole, O.J.; Tallon, S.J.; Grey, J.B.; Fletcher, K.; Fletcher, A.J. Extraction of lipids from a specialist dairy stream. *J. Supercritic. Fluids* **2008**, *45*, 314–321. [CrossRef]

98. Astaire, J.C.; Ward, R.; German, J.B.; Jiménez-Flores, R. Concentration of polar MFGM lipids from buttermilk by microfiltration and supercritical fluid extraction. *J. Dairy Sci.* **2003**, *86*, 2297–2307. [CrossRef]

99. Tomasula, P.M.; Bonnaillie, L.M. Crossflow microfiltration in the dairy industry. In *Emerging Dairy Processing Technologies*, 1st ed.; Datta, N., Tomasula, P., Eds.; John Wiley & Sons, Ltd.: Oxford, UK, 2015; pp. 1–32.

100. Pimentel, L.; Gomes, A.; Pintado, M.; Rodríguez-Alcalá, L.M. Isolation and analysis of phospholipids in dairy foods. *J. Analytic. Methods Chem.* **2016**, *e9827369*, 1–12. [CrossRef] [PubMed]

101. Kim, S.-H.; Min, C.-S. Fouling reduction using the resonance vibration in membrane separation of whole milk. *J. Indust. Eng. Chem.* **2019**, *75*, 123–129. [CrossRef]

102. Sachdeva, S.; Buchheim, W. Recovery of phospholipids from buttermilk using membrane processing. *Kieler Milchwirtsch. Forsch.* **1997**, *49*, 47–68.

103. ISO 14044:2006(en) Environmental Management—Life Cycle Assessment—Requirements and Guidelines. Available online: https://www.iso.org/obp/ui/#iso:std:iso:14044:ed-1:v1:en (accessed on 27 February 2020).

104. Famiglietti, J.; Guerci, M.; Proserpio, C.; Ravaglia, P.; Motta, M. Development and testing of the product environmental footprint milk tool: A comprehensive LCA tool for dairy products. *Sci. Total Environ.* **2019**, *648*, 1614–1626. [CrossRef] [PubMed]

105. International Standard Organization. *Greenhouse Gases—Carbon Footprint of Products—Requirements and Guidelines for Quantification*; International Standard Organization: Geneva, Switzerland, 2018; pp. 1–11.

106. Intergovernmental Panel on Climate Change. *Guidelines for National Greenhouse Gas Inventories. Volume 4: Agriculture, Forestry and Other Land Use*; Intergovernmental Panel on Climate Change: New York, NY, USA, 2006; p. 132.

107. Verge, X.P.; Dyer, J.A.; Worth, D.E.; Smith, W.N.; Desjardins, R.L.; McConkey, B.G. A greenhouse gas and soil carbon model for estimating the carbon footprint of livestock production in Canada. *Animals (Basel)* **2012**, *2*, 437–454. [CrossRef] [PubMed]

108. Dyer, J.A.; Desjardins, R.L. Simulated farm fieldwork, energy consumption and related greenhouse gas emissions in Canada. *Biosys. Eng.* **2003**, *85*, 503–513. [CrossRef]

109. Feitz, A.J.; Lundie, S.; Dennien, G.; Morain, M.; Jones, M. Generation of an industry-specific physico-chemical allocation matrix. Application in the dairy industry and implications for systems analysis (9 pp). *Int. J. Life Cycle Assess.* **2005**, *12*, 109–117. [CrossRef]

Foods **2020**, *9*, 263

110. Mercier-Bouchard, D.; Benoit, S.; Doyen, A.; Britten, M.; Pouliot, Y. Process efficiency of casein separation from milk using polymeric spiral-wound microfiltration membranes. *J. Dairy Sci.* **2017**, *100*, 8838–8848. [CrossRef]

111. Ang, B.W.; Su, B. Carbon emission intensity in electricity production: A global analysis. *Energy Policy* **2016**, *94*, 56–63. [CrossRef]

112. Azapagic, A. *CCaLC BIOCHEM V3.0 Carbon Footprint Calculator (Database of Ecoinvent 3), Funded by the Carbon Trust, EPSRC and NERC (Grant Number EP/F003501/1)*; The University of Manchester: Manchester, UK, 9 January 2013.

113. Matzen, M.; Demirel, Y. Methanol and dimethyl ether from renewable hydrogen and carbon dioxide: Alternative fuels production and life-cycle assessment. *J. Clean. Prod.* **2016**, *139*, 1068–1077. [CrossRef]

114. Albarelli, J.Q.; Santos, D.T.; Cocero, M.J.; Meireles, M.A.A. Perspectives on the integration of a supercritical fluid extraction plant to a sugarcane biorefinery: Thermo-economical evaluation of CO2 recycle systems. *Food Sci. Technol.* **2017**, *38*, 13–18. [CrossRef]

115. Natolino, A. Application of Supercritical Fluids Technology on Winery by-Products. Ph.D. Thesis, University of Udine, Udine, Italy, 2016.

116. Methot-Hains, S.; Benoit, S.; Bouchard, C.; Doyen, A.; Bazinet, L.; Pouliot, Y. Effect of transmembrane pressure control on energy efficiency during skim milk concentration by ultrafiltration at 10 and 50 degrees C. *J. Dairy Sci.* **2016**, *99*, 8655–8664. [CrossRef]

117. Zhao, R.; Xu, Y.; Wen, X.; Zhang, N.; Cai, J. Carbon footprint assessment for a local branded pure milk product: A lifecycle based approach. *Food Sci. Technol.* **2017**, *38*, 98–105. [CrossRef]

118. Berlina, J. Environmental life cycle assessment (LCA) of Swedish semi-hard cheese. *Int. Dairy J.* **2002**, *12*, 939–953. [CrossRef]

MDPI
St. Alban-Anlage 66
4052 Basel
Switzerland
Tel. +41 61 683 77 34
Fax +41 61 302 89 18
www.mdpi.com

Foods Editorial Office
E-mail: foods@mdpi.com
www.mdpi.com/journal/foods